COREQUISITE RESOURCE NOTEBOOK FOR COLLEGE ALGEBRA

JUDITH A. BEECHER   JUDITH A. PENNA

COLLEGE ALGEBRA

FIFTH EDITION

Judith A. Beecher

Judith A. Penna

Marvin L. Bittinger

Indiana University Purdue University Indianapolis

P Pearson

P Pearson

ISBN-13: 978-0-13-529939-5
ISBN-10: 0-13-529939-X

Contents

Student Activities 141

A Note from the Authors

As we prepared to write this Notebook we asked instructors what resources they would like to have available for a corequisite or support course to accompany our textbook, *College Algebra*, 5th Edition, by Beecher, Penna, and Bittinger, and also what resources would be of most value to students. Our goal is to give students the best possible chance to understand the material and to be successful in their College Algebra course by providing instructors with the highest quality resources available. With this goal in mind, we have prepared a multifaceted Notebook with a variety of resources that can be used in flexible and creative ways.

In preparing these materials, we had one-on-one conversations with instructors who are leaders in the corequisite movement. It became clear that "co-requisite" and/or "support" mean different things at different institutions. What we have tried to do is to present thoughtfully prepared materials that instructors can use in many different ways depending on what the structure of such a course is at their particular institution.

The Notebook itself, which is available in print and within MyLab Math, consists of three elements: Integrated Review Worksheets, Interactive Preview Worksheets, and Student Activities. In addition, videos and guided visualizations are available within MyLab Math. Each will be explained in more detail below.

Although the topics in the worksheets and the activities are presented in the order in which they occur in the *College Algebra* textbook, instructors could easily choose to teach them in a different order if that is their preference. They could also choose to use a Review Worksheet to preview a topic or to use a Preview Worksheet to review a topic.

Guides correlating the Review Worksheets, the Preview Worksheets, the Activities, and the Guided Visualizations to the *College Algebra* textbook are placed near the front of the Notebook. Individual correlation guides are also found at the beginning of each section of material.

Integrated Review Worksheets

There are 23 worksheets that review Intermediate Algebra coverage of topics that will be presented in the College Algebra course. Although it is true that many of the topics that are taught in College Algebra are also taught in an Intermediate Algebra course, students will have greater success if some of these are reviewed before they are presented in College Algebra.

The Review Worksheets are intended to be used in a classroom setting with an instructor who can offer additional explanations and answer questions when needed. Each worksheet is intended to be filled out and completed in a single class session. All of the answers to the exercises are included in the Notebook so that students can check their work immediately.

The filled-out worksheets will be a valuable resource for study and review. Each Integrated Review Worksheet is also accompanied by a video that gives additional explanation and examples.

Interactive Preview Worksheets

The 16 Preview Worksheets provide a preview of a topic to be taught in College Algebra. Some of these will focus on a particular aspect of a topic that will provide a non-intimidating introduction to that topic. Others provide an intuitive approach to a concept. Many of these worksheets make use of visual elements such as graphs to give a student a view of the "big picture." The Guided Visualizations can also be used in conjunction with the Preview Worksheets as an additional visual element. We find that incorporating visual elements wherever possible gives a boost in understanding.

The Preview Worksheets are structured much like the Review Worksheets, with examples and with exercises that are intended to be completed in a single class session. All of the answers to the exercises are provided in the Notebook.

Activities

There are 19 activities in the Notebook, with at least one for each chapter of the *College Algebra* text. Many of these are group activities. As instructors know, when students work together the opportunities to discuss concepts and to explain their reasoning to other students serve to increase understanding.

Videos

As mentioned above, each of the 23 Interactive Review Worksheets is accompanied by a video that gives additional instruction and examples.

Guided Visualizations

These animated figures help students visualize concepts through directed explorations and purposeful manipulation. They encourage active learning, critical thinking, and conceptualization. Located in the Video & Resource Library within MyLab Math, these can also be assigned as homework with correlating assessment exercises. Additional Exploratory Activities are available to help students think more conceptually about the figures and provide an excellent framework for group projects or discussions. They can be used along with the Review Worksheets and Preview Worksheets or as a stand-alone element.

In conclusion, we want to emphasize that we want this Notebook to contribute to student understanding and success and to provide instructors with resources that are of value to them as they guide their students to that goal.

Best wishes to all,
Judy Beecher
Judy Penna

Correlation Guides

The **Review and Preview Worksheets** in this Notebook accompany *College Algebra*, 5th edition, by Beecher/Penna/Bittinger. The following table contains a correlation between the sections in the text and the worksheets.

Section in the Text	Review Worksheet #	Preview Worksheet #
1.1	1, 2, 3	
1.2	2, 3, 4, 5, 6, 7	1
1.3	1, 2	
1.4	2, 8	
1.5	8	
1.6	7, 9	
2.1	7, 10, 11	2
2.2	11, 12, 13	
2.3	7, 13	
2.5		3
3.1	14, 15, 16	
3.2	4, 5, 6, 15, 17	
3.3	4, 5, 17	4
3.4	4, 5, 6, 18, 19	5, 6
3.5		7
4.1	20, 21	8
4.3	2	
4.5	7	9
4.6	7	
5.1		10
5.2	22	
5.3	23	
5.5		11, 12
6.1	1, 2	
6.2	2	
6.3	2	
6.4	2	
6.5	2	13
7.2	17	15
7.3		14, 16

The **Student Activities** in this Notebook accompany *College Algebra*, 5th edition, by Beecher/Penna/Bittinger. The following table contains a correlation between the chapter in the text and the activity.

Chapter in the Text	Student Activity #
Just-In-Time 4	1
Just-In-Time 10	2
Just-In-Time 13	3
Just-In-Time 13	4
Chapter 1	5
Chapter 1	6
Chapter 1	7
Chapter 2	8
Chapter 2	9
Chapter 3	10
Chapter 3	11
Chapter 3	12
Chapter 4	13
Chapter 5	14
Chapter 6	15
Chapter 6	16
Chapter 6	17
Chapter 7	18
Chapter 8	19

The **Guided Visualizations** listed below accompany *College Algebra*, 5th edition, by Beecher/Penna/Bittinger. They are located in the Video and Resource Library within MyLab Math. The following table correlates these figures with the text and the Preview Worksheets.

Guided Visualization	Section in Text	Preview Worksheet
Graphing Linear Equations	1.1	
The Distance Formula	1.1	
Equations of Circles	1.1	Preview #14
Domain and Range of Functions	1.2	Preview #1
Graphing Functions	1.2	Preview #1
Slope	1.3	
Slope Between Two Points	1.3	
Slope-Intercept Form	1.3	
Equations of Lines: Slope-Intercept Form	1.3	
Equations of Lines: Point-Slope Form	1.4	
Sum and Difference of Two Functions	2.2	
Product and Quotient of Two Functions	2.2	
Difference Quotients	2.2	
Symmetry of Functions: Even and Odd	2.4	Preview #3
Intercepts and Solutions	3.2	Preview #8
Quadratic Functions and Their Graphs	3.3	Preview #4
Graphs of Quadratic Functions	3.3	Preview #4
Application: Height of a Baseball	3.3	Preview #4
Absolute-Value Equations and Inequalities	3.5	Preview #7
Leading-Term Test and Polynomials	4.1	
Zeros of Polynomial Functions	4.1	Preview #8
Application: Volume of a Box	4.1	
The Intermediate-Value Theorem	4.2	
Graphs of Rational Functions Part I	4.5	Preview #9
Graphs of Rational Functions Part II	4.5	Preview #9
Graphs of Rational Functions: Oblique Asymptotes	4.5	
Polynomial and Rational Inequalities	4.6	
Graphing Functions and Their Inverses	5.1	Preview #10
Graphs of Inverse Functions	5.1	Preview #10
Graphs of Exponential Functions	5.2	
Graphs of Logarithmic Functions	5.3	
Exponential Growth Models	5.6	
Graphs of Logistic Functions	5.6	
Mixture Problems	6.1	
Motion Problems	6.1	
Linear Inequalities in Two Variables	6.7	
Graphing Hyperbolas	7.3	Preview #16
Geometric Sequences and Series	8.3	

Integrated Review Worksheets

Correlation Guide

The **Review Worksheets** in this Notebook accompany *College Algebra*, 5th edition, by Beecher/Penna/Bittinger. The following table contains a correlation between the worksheets and the sections in the text.

Review Worksheet #	Sections in the Text
1	1.1, 1.3, 6.1
2	1.1-1.4, 4.3, 6.1- 6.5
3	1.1, 1.2
4	1.2, 3.2-3.4
5	1.2, 3.2-3.4
6	1.2, 3.2, 3.4
7	1.2, 1.6, 2.1, 2.3, 4.5, 4.6
8	1.4, 1.5
9	1.6
10	2.1
11	2.1, 2.2
12	2.2
13	2.2, 2.3
14	3.1
15	3.1, 3.2
16	3.1
17	3.2, 3.3, 7.2
18	3.4
19	3.4
20	4.1
21	4.1
22	5.2
23	5.3

Integrated Review 1: GRAPHING LINEAR EQUATIONS

Equations of the form $y = mx + b$, $Ax + By = C$, $x = a$ and $y = b$ are said to be *linear* because the graph of each equation is a straight line. The following equations are examples of linear equations.

$$y = 7x - \frac{3}{4}, \qquad 3x + 9y = 20, \qquad x = -5, \qquad y = 8$$

The graph of an equation is a drawing that represents all of its solutions. The solutions are ordered pairs with coordinates listed in alphabetical order of the variables.

Linear Equations of the Form $y = mx + b$

Example 1 Graph $y = 2x + 3$.

If $x = -2$, $y = 2(-2) + 3 = -4 + 3 = -1$.

If $x = 0$, $y = 2(0) + 3 = 0 + 3 = 3$.

If $x = 1$, $y = 2(1) + 3 = 2 + 3 = 5$.

x	y
-2	-1
0	3
1	5

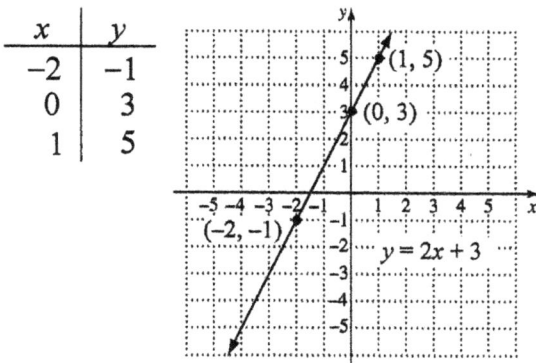

Example 2 Graph $y = -\frac{2}{5}x$.

If $x = -5$, $y = -\frac{2}{5}(-5) = 2$.

If $x = 0$, $y = -\frac{2}{5}(0) = 0$.

If $x = 5$, $y = -\frac{2}{5}(5) = -2$.

x	y
-5	2
0	0
5	-2

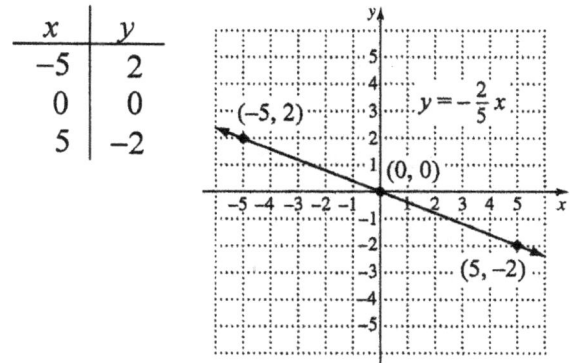

Linear Equations of the Form $Ax + By = C$

Example 3 Graph $4y + 3x = -8$.

We first solve for y to find an equivalent form $y = mx + b$.

	$4y + 3x = -8$
Subtract $3x$ on both sides.	$4y = -3x - 8$
Multiply by $\frac{1}{4}$ on both sides.	$\frac{1}{4}(4y) = \frac{1}{4}(-3x - 8)$
Simplify.	$y = -\frac{3}{4}x - 2$

If $x = 4$, $y = -\frac{3}{4}(4) - 2 = -3 - 2 = -5$.

If $x = 0$, $y = -\frac{3}{4}(0) - 2 = 0 - 2 = -2$.

If $x = -4$, $y = -\frac{3}{4}(-4) - 2 = 3 - 2 = 1$.

x	y
4	-5
0	-2
-4	1

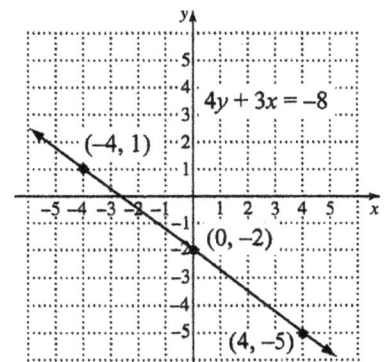

When the coefficients of the x and y-terms in $Ax + By = C$ are factors of the constant C, it is easier to graph using intercepts.

Intercepts

The **y-intercept** is $(0, b)$. To find b, let $x = 0$ and solve the equation for y.

The **x-intercept** is $(a, 0)$. To find a, let $y = 0$ and solve the equation for x.

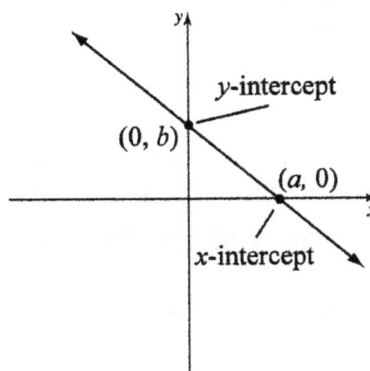

Example 4 Graph $-2x + 3y = 6$.

To find the y-intercept, we let $x = 0$ and solve for y.

$$-2x + 3y = 6$$
$$-2(0) + 3y = 6$$
$$3y = 6$$
$$y = 2$$

$(0, 2)$ is the y-intercept.

To find the x-intercept, we let $y = 0$ and solve for x.

$$-2x + 3y = 6$$
$$-2x + 3(0) = 6$$
$$-2x = 6$$
$$x = -3$$

$(-3, 0)$ is the x-intercept.

Find a third point as a check. Here we let $x = 3$.

$$-2x + 3y = 6$$
$$-2(3) + 3y = 6$$
$$-6 + 3y = 6$$
$$3y = 12$$
$$y = 4$$

$(3, 4)$ is also on the graph.

x	y
0	2
-3	0
3	4

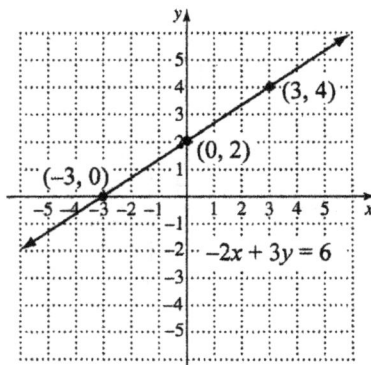

Linear Equations of the Form $x = a$ and $y = b$

Example 5 Graph $y = -4$.

Think of $y = -4$ as $0 \cdot x + y = -4$.

x	y
-2	-4
0	-4
3	-4

x can be any number

y must be -4.

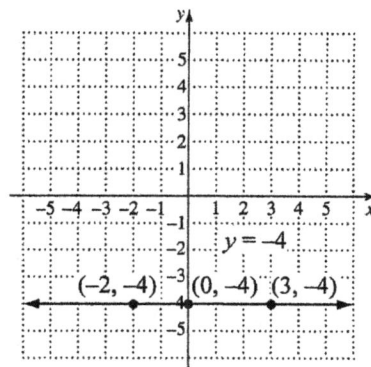

Example 6 Graph $x = 3$.

Think of $x = 3$ as $x + 0 \cdot y = 3$.

x	y
3	-4
3	0
3	5

x must be 3.

y can be any number.

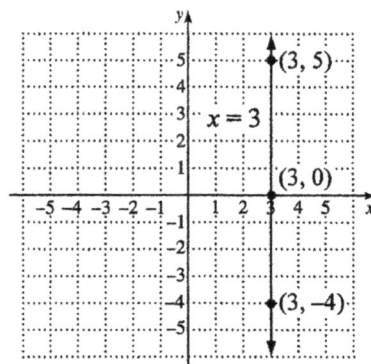

Horizontal Lines and Vertical Lines

The graph of $y = b$ is a horizontal line. The y-intercept is $(0, b)$.
The graph of $x = a$ is a vertical line. The x-intercept is $(a, 0)$.

Check Your Understanding

Fill-in the coordinates of each intercept.

1.
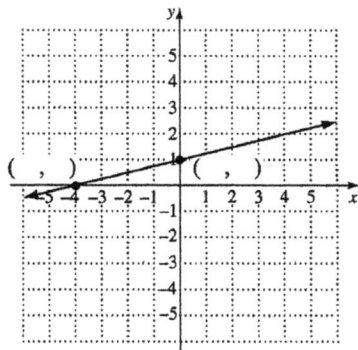

2.
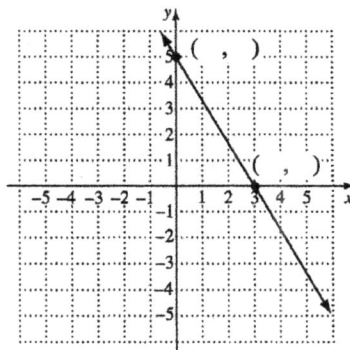

3.
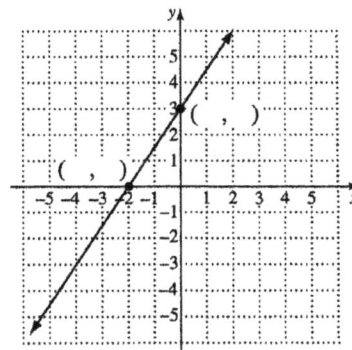

Determine whether the statement is true or false.

4. To find the x-intercept of the graph of $2x - 7y = -14$, let $x = 0$.

5. The second coordinate of each point of the graph of $y = -1$ is -1.

6. The graph of a linear equation is always a straight line.

Match each equation with its graph from choices A – D.

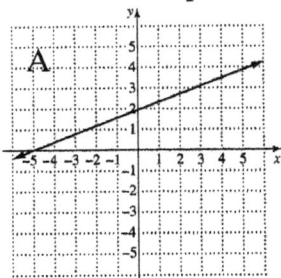

A

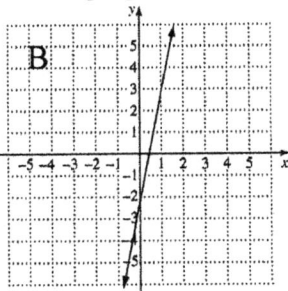

B

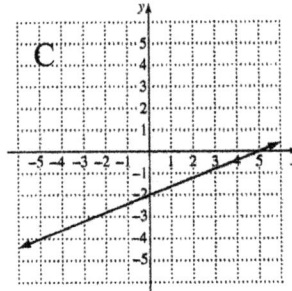

C

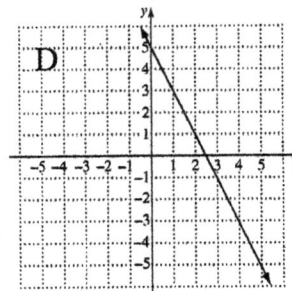

D

7. $2x - 5y = 10$

x	y

8. $y = 5x - 2$

x	y

9. $y = -2x + 5$

x	y

10. $5y - 2x = 10$

x	y

Exercises
Graph.

1. $y = -x$

x	y

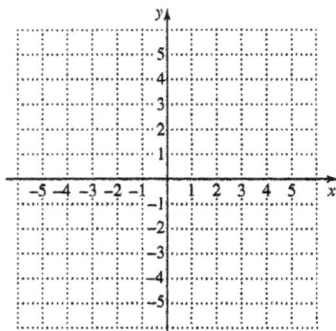

2. $y = -4x + 5$

x	y

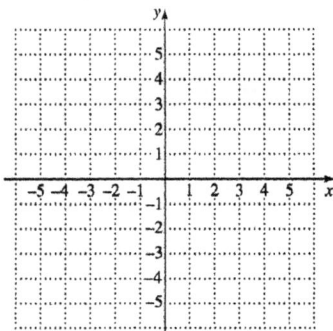

3. $2x - 3y = 9$

x	y

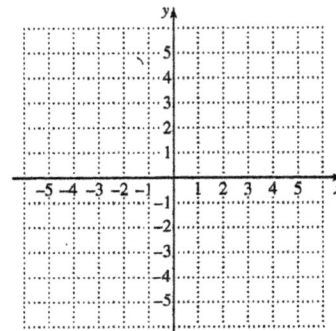

4. $5x - 4y = 20$

5. $x = -1$

6. $3x - 5y = 10$

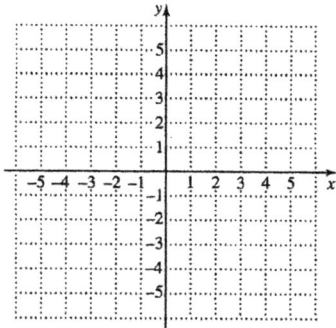

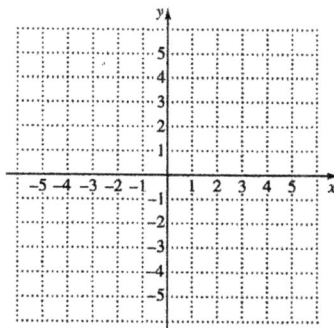

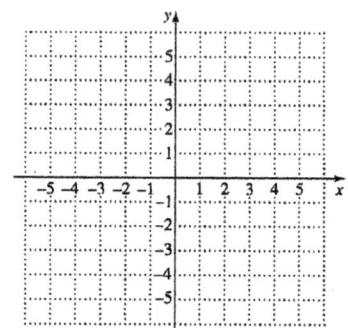

7. $y = \dfrac{2}{3}x - 4$

8. $x + y = 5$

9. $3x - 2y = 6$

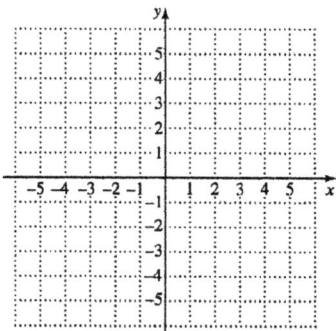

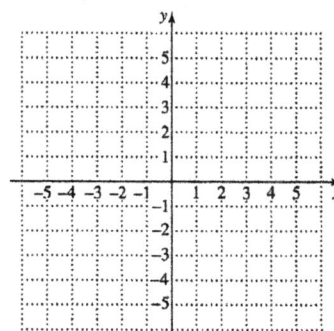

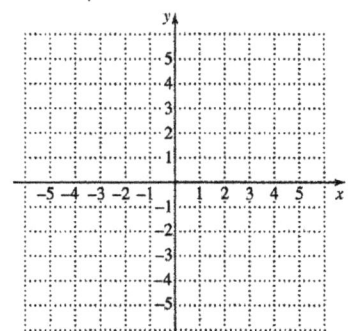

10. $y = 2$

11. $y + \dfrac{1}{2}x = 1$

12. $3x + y = 7$

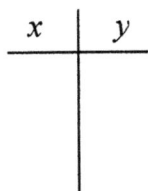

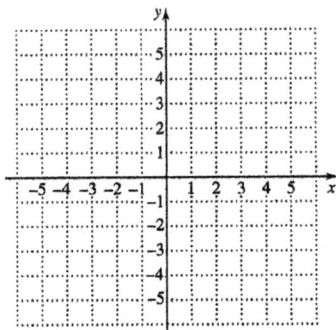

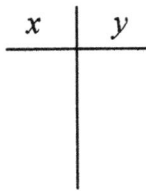

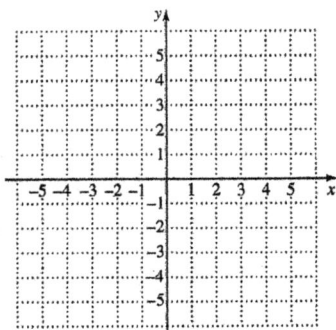

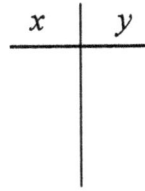

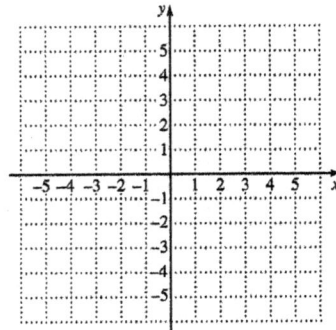

Notes:

Integrated Review 2: OPERATIONS ON THE REAL NUMBERS

Adding Real Numbers

1. *Positive numbers*: Add the numbers. The result is positive.
2. *Negative numbers*: Add absolute values. Make the answer negative.
3. *A positive number and a negative number*:
   - If the numbers have the same absolute value, the answer is 0.
   - If the numbers have different absolute values, subtract the smaller absolute value from the larger. Then:
     a) If the positive number has the greater absolute value, make the answer positive.
     b) If the negative number has the greater absolute value, make the answer negative.
4. *One number is zero*: The sum is the other number.

Examples Add.

1. $5 + 8 = 13$

2. $-5 + 8 = 3$

3. $5 + (-8) = -3$

4. $-5 + (-8) = -13$

5. $-8 + 0 = -8$

6. $21 + (-7) = 14$

7. $-21 + (-7) = -28$

8. $-23 + 23 = 0$

9. $6.2 + (-17.1) = -10.9$

10. $-\dfrac{5}{6} + \left(-\dfrac{5}{8}\right) = -\dfrac{5}{6} \cdot \dfrac{4}{4} + \left(-\dfrac{5}{8}\right) \cdot \dfrac{3}{3}$

$\qquad\qquad = -\dfrac{20}{24} + \left(-\dfrac{15}{24}\right) = -\dfrac{35}{24}$

Subtracting Real Numbers

For any real numbers a and b, $a - b = a + (-b)$.

We can subtract by adding the **opposite (additive inverse)** of the number being subtracted. (Two numbers whose sum is 0 are called opposites, or additive inverses. For example, the opposite of -2 is 2 and the opposite of $\dfrac{4}{7}$ is $-\dfrac{4}{7}$.)

Examples Subtract. Rewrite each subtraction as an addition and then use the rules for adding real numbers.

11. $24 - 9 = 24 + (-9) = 15$

12. $24 - (-9) = 24 + 9 = 33$

13. $-24 - 9 = -24 + (-9) = -33$

14. $-24 - (-9) = -24 + 9 = -15$

15. $0 - 9 = 0 + (-9) = -9$

16. $0.13 - 0.7 = 0.13 + (-0.7) = -0.57$

17. $\dfrac{2}{3} - \left(-\dfrac{5}{3}\right) = \dfrac{2}{3} + \dfrac{5}{3} = \dfrac{7}{3}$

18. $-4.3 - 10 = -4.3 + (-10) = -14.3$

19. $-\dfrac{7}{8} - \left(-\dfrac{2}{5}\right) = -\dfrac{7}{8} + \dfrac{2}{5}$

$= -\dfrac{7}{8} \cdot \dfrac{5}{5} + \dfrac{2}{5} \cdot \dfrac{8}{8}$

$= -\dfrac{35}{40} + \dfrac{16}{40} = -\dfrac{19}{40}$

20. $\dfrac{11}{60} - \dfrac{5}{24} = \dfrac{11}{60} + \left(-\dfrac{5}{24}\right) = \dfrac{11}{60} \cdot \dfrac{2}{2} + \left(-\dfrac{5}{24}\right) \cdot \dfrac{5}{5}$

$= \dfrac{22}{120} + \left(-\dfrac{25}{120}\right) = -\dfrac{3}{120}$

$= -\dfrac{1 \cdot 3}{40 \cdot 3} = -\dfrac{1}{40} \cdot \dfrac{3}{3} = -\dfrac{1}{40} \cdot 1 = -\dfrac{1}{40}$

Multiplying or Dividing Real Numbers

1. Multiply or divide the absolute values.
2. If the signs are the same, then the answer is positive.
3. If the signs are different, then the answer is negative.

Examples Multiply.

21. $-3 \times 10 = -30$

22. $-3 \times (-10) = 30$

23. $\dfrac{7}{11} \cdot \dfrac{9}{2} = \dfrac{63}{22}$

24. $0.5 \times (-1.2) = -0.6$

25. $400(-300) = -120,000$

26. $-\dfrac{5}{8} \cdot \dfrac{3}{20} = -\dfrac{15}{160} = -\dfrac{5 \cdot 3}{5 \cdot 32}$

$$= \dfrac{5}{5} \cdot \left(-\dfrac{3}{32}\right) = 1 \cdot \left(-\dfrac{3}{32}\right) = -\dfrac{3}{32}$$

Examples Divide.

27. $\dfrac{-42}{-6} = 7$

28. $500 \div (-50) = -10$

29. $\dfrac{-33}{11} = -3$

30. $-1.44 \div (-1.2) = 1.2$

31. $-\dfrac{1}{3} \div 3 = -\dfrac{1}{3} \cdot \dfrac{1}{3} = -\dfrac{1}{9}$

32. $-2 \div \left(-\dfrac{3}{4}\right) = -\dfrac{2}{1} \cdot \left(-\dfrac{4}{3}\right) = \dfrac{8}{3}$

33. $\dfrac{4}{15} \div \left(-\dfrac{3}{10}\right) = \dfrac{4}{15} \cdot \left(-\dfrac{10}{3}\right)$

$$= -\dfrac{40}{45} = -\dfrac{5 \cdot 8}{5 \cdot 9}$$

$$= \dfrac{5}{5} \cdot \left(-\dfrac{8}{9}\right) = 1 \cdot \left(-\dfrac{8}{9}\right) = -\dfrac{8}{9}$$

Check Your Understanding

Match each expression with an equivalent expression from choices A – H.

1. $-10 - (-31)$

2. $-\dfrac{1}{4} \div \left(-\dfrac{2}{5}\right)$

   A. $-10 + (-31)$

   E. $-\dfrac{1}{4} \cdot \left(-\dfrac{5}{2}\right)$

3. $-\dfrac{1}{4} \div 5$

4. $10 - 31$

   B. $10 + (-31)$

   F. $4 \cdot \dfrac{5}{2}$

5. $\dfrac{2}{5} \div (-4)$

6. $10 - (-31)$

   C. $-10 + 31$

   G. $\dfrac{2}{5} \cdot \left(-\dfrac{1}{4}\right)$

7. $-10 - 31$

8. $4 \div \dfrac{2}{5}$

   D. $10 + 31$

   H. $-\dfrac{1}{4} \cdot \dfrac{1}{5}$

Exercises Perform the indicated operation.

1. $-3+17$

2. $-11-(-6)$

3. $-80\cdot(-3)$

4. $\dfrac{42}{-21}$

5. $\dfrac{1}{4}-\dfrac{5}{2}$

6. $-\dfrac{2}{5}\div 5$

7. $-5.14-(-0.3)$

8. $0-8$

9. $-24\div(-3)$

10. $-9+(-25)$

11. $-21\div(0.3)$

12. $-\dfrac{4}{9}\cdot\dfrac{5}{2}$

13. $-3\div\left(-\dfrac{11}{12}\right)$

14. $4.5(-1.2)$

15. $\dfrac{15}{8}-\left(-\dfrac{5}{3}\right)$

16. $\dfrac{1}{4}\div(-4)$

17. $-15\div 12$

18. $-\dfrac{14}{9}\div\left(-\dfrac{7}{3}\right)$

19. $3.16+(-5.2)$

20. $8-(-20)$

Integrated Review 3: ORDER OF OPERATIONS

Rules for Order of Operations

1. Do all the calculations within grouping symbols, like parentheses, before operations outside.
2. Evaluate all exponential expressions.
3. Do all multiplications and divisions in order from left to right.
4. Do all additions and subtractions in order from left to right.

Example 1 Simplify: $9 + 4(7 - 1)$.

Subtract $7 - 1$.	$= 9 + 4(6)$
Multiply $4(6)$.	$= 9 + 24$
Add $9 + 24$.	$= 33$

Example 2 Simplify: $-48 \div (-6) \div (-2)$.

Divide $-48 \div (-6)$.	$= 8 \div (-2)$
Divide $8 \div (-2)$.	$= -4$

Example 3 Simplify: $-3 \cdot 15 + 25 \div 5$.

Multiply $-3 \cdot 15$.	$= -45 + 25 \div 5$
Divide $25 \div 5$.	$= -45 + 5$
Add $-45 + 5$.	$= -40$

Example 4 Simplify: $24 - 16 \div 4$.

Divide $16 \div 4$.	$= 24 - 4$
Subtract $24 - 4$.	$= 20$

Example 5 Simplify: $8 \cdot 11 - 3^2 - 2^3$.

Evaluate 3^2.	$= 8 \cdot 11 - 9 - 2^3$
Evaluate 2^3.	$= 8 \cdot 11 - 9 - 8$
Multiply $8 \cdot 11$.	$= 88 - 9 - 8$
Subtract $88 - 9$.	$= 79 - 8$
Subtract $79 - 8$.	$= 71$

Example 6 Simplify: $27 + (9 - 12)^2 \div 3$.

Subtract $9 - 12$.	$= 27 + (-3)^2 \div 3$
Evaluate $(-3)^2$.	$= 27 + 9 \div 3$
Divide $9 \div 3$.	$= 27 + 3$
Add $27 + 3$.	$= 30$

Example 7 Simplify: $20 - 5(10 \div 2) + 5 \cdot 50$.

Divide $10 \div 2$.	$= 20 - 5(5) + 5 \cdot 50$
Multiply $5(5)$.	$= 20 - 25 + 5 \cdot 50$
Multiply $5 \cdot 50$.	$= 20 - 25 + 250$
Subtract $20 - 25$.	$= -5 + 250$
Add $-5 + 250$.	$= 245$

Example 8 Simplify: $9 \cdot 10^2 - 4^3 + 40 \div 4$.

Evaluate 10^2.	$= 9 \cdot 100 - 4^3 + 40 \div 4$
Evaluate 4^3.	$= 9 \cdot 100 - 64 + 40 \div 4$
Multiply $9 \cdot 100$.	$= 900 - 64 + 40 \div 4$
Divide $40 \div 4$.	$= 900 - 64 + 10$
Subtract $900 - 64$.	$= 836 + 10$
Add $836 + 10$.	$= 846$

Check Your Understanding

Select the correct step-by-step solution. Explain the error in the incorrect solution.

1. Simplify: $18 \div 6 + 3 - 5$.

   A. $18 \div 6 + 3 - 5$
   $= 3 + 3 - 5$ (1)
   $= 6 - 5$ (2)
   $= 1$ (3)

   B. $18 \div 6 + 3 - 5$
   $= 18 \div 9 - 5$ (1)
   $= 2 - 5$ (2)
   $= -3$ (3)

2. Simplify: $80 - 4^2 \cdot 5 \div 5(10 - 2)$.

   A. $80 - 4^2 \cdot 5 \div 5(10 - 2)$
   $= 80 - 4^2 \cdot 5 \div 5(8)$ (1)
   $= 80 - 16 \cdot 5 \div 5(8)$ (2)
   $= 80 - 80 \div 5(8)$ (3)
   $= 80 - 80 \div 40$ (4)
   $= 80 - 2$ (5)
   $= 78$ (6)

   B. $80 - 4^2 \cdot 5 \div 5(10 - 2)$
   $= 80 - 4^2 \cdot 5 \div 5(8)$ (1)
   $= 80 - 16 \cdot 5 \div 5(8)$ (2)
   $= 80 - 80 \div 5(8)$ (3)
   $= 80 - 16(8)$ (4)
   $= 80 - 128$ (5)
   $= -48$ (6)

Exercises In Exercises 1 and 2, describe the calculation in each step of the solution.

1. Simplify: $6 - 3 \cdot 5 + 8 \div 2^2$.

_____	$= 6 - 3 \cdot 5 + 8 \div 4$
_____	$= 6 - 15 + 8 \div 4$
_____	$= 6 - 15 + 2$
_____	$= -9 + 2$
_____	$= -7$

2. Simplify: $32 \div 4 \cdot 2^2 - 2(5 - 3)$.

_____	$= 32 \div 4 \cdot 2^2 - 2 \cdot 2$
_____	$= 32 \div 4 \cdot 4 - 2 \cdot 2$
_____	$= 8 \cdot 4 - 2 \cdot 2$
_____	$= 32 - 2 \cdot 2$
_____	$= 32 - 4$
_____	$= 28$

Simplify.

3. $9 - 2(7 + 4)$

4. $9 \cdot 8 - 2 \cdot 6$

5. $-2 + 2 \cdot 3$

6. $300 \div 15 \div 5$

7. $8 \div 2(5 - 3)^2$

8. $3 \cdot 8^2 - 5^2 \div 5$

9. $-12 - 6 \div 2 - (-4)$

10. $35 - 5(14 \div 2) + 7 \cdot 10$

11. $24 - (2-4)^2 \div 2$

12. $4 \cdot \left(\dfrac{1}{2}\right)^2 - (3-8)$

13. $\dfrac{1}{3}(12-3)^2$

14. $6 \cdot 7^2 + 2^3 - 50 \div 5$

15. $6^3 + 25 \cdot 6 \div \left(20 + \dfrac{1}{2} \cdot 10\right)$

16. $8(7-1) \div 2 - 3(-2)^2$

17. $-81 + 9 \cdot 3 \div (-3)^2 - 27$

18. $5 \cdot 5^2 - 5 \div 5 + 5(5-2)$

19. $\left[2(11-8)\right]^2$

20. $\left[18 \div (2+1)\right]^3$

Integrated Review 4: FACTORING TRINOMIALS: $ax^2 + bx + c$

Let's begin by reviewing the FOIL method of multiplying two binomials.

$$\text{Multiply: } (x+p)(x+q) = \overset{F}{x^2} + \overset{O}{qx} + \overset{I}{px} + \overset{L}{pq}$$

$$= x^2 + (p+q)x + pq$$

Example: $(x+4)(x+7) = x^2 + 7x + 4x + 4 \cdot 7$

$$= x^2 + (4+7)x + 28 = x^2 + 11x + 28$$

To factor a trinomial, we think of FOIL in reverse.

Factor: $x^2 + (p+q)x + pq = (x+p)(x+q)$

Example: $x^2 + 11x + 28 = x^2 + (4+7)x + 4 \cdot 7$

$$= (x+4)(x+7)$$

We look for two numbers whose product is 28 and whose sum is 11. Those numbers are 4 and 7.

Example 1 Factor: $x^2 + 8x + 15$.

The first term is x^2, so the first term in each binomial is x.

$x^2 + 8x + 15 = (x\quad)(x\quad)$

Constant, 15, is *positive*; coefficient of middle term, 8, is *positive*.

Find two numbers whose product is 15 and whose sum is 8.

Pairs of factors	Sums of factors
1, 15	16
$-1, -15$	-16
3, 5	8 $\leftarrow$
$-3, -5$	-8

The numbers we need are 3 and 5.

$x^2 + 8x + 15 = (x+3)(x+5)$

Example 2 Factor: $x^2 - 8x + 15$.

The first term is x^2, so the first term in each binomial is x.

$x^2 - 8x + 15 = (x\quad)(x\quad)$

Constant, 15, is *positive*; coefficient of middle term, -8, is *negative*.

Find two numbers whose product is 15 and whose sum is -8. (See the table in Example 1.)

The numbers we need are -3 and -5.

$x^2 - 8x + 15 = (x-3)(x-5)$

When the constant term of a trinomial is positive, we look for two factors with the same sign (both positive or both negative). The sign is that of the middle term.

Example 3 Factor: $y^2 + 9y - 36$.

The first term in each binomial is y.

$$y^2 + 9y - 36 = (y \quad)(y \quad)$$

Constant, -36, is *negative*; coefficient of middle term, 9, is *positive*.

Find two numbers, whose product is -36 and whose sum is 9.

Pairs of factors	Sums of factors
1, -36	-35
-1, 36	35
2, -18	-16
-2, 18	16
3, -12	-9
-3, 12	9 ←
4, -9	-5
-4, 9	5
6, -6	0

The numbers we need are -3 and 12.

$$y^2 + 9y - 36 = (y - 3)(y + 12)$$

Example 4 Factor: $y^2 - 9y - 36$.

The first term in each binomial is y.

$$y^2 - 9y - 36 = (y \quad)(y \quad)$$

Constant, -36, is *negative*; coefficient of middle term, -9, is *negative*.

Find two numbers whose product is -36 and whose sum is -9. (See the table in Example 3.) The numbers we need are 3 and -12.

$$y^2 - 9y - 36 = (y + 3)(y - 12)$$

When the constant term of a trinomial is negative, we look for two factors whose product is negative. One of them must be positive and the other negative. Their sum must be the coefficient of the middle term.

Exercises Factor.

1. $x^2 + 5x - 36$

2. $y^2 - 10y + 9$

3. $y^2 - 3y - 10$

4. $s^2 - 9s - 90$

5. $x^2 + 7x - 8$

6. $b^2 + 8b + 12$

7. $x^2 - x - 56$

8. $a^2 - 3a - 54$

9. $y^2 - 5y - 14$

10. $w^2 - 10w - 11$

11. $q^2 + 19q + 90$

12. $x^2 - 14x + 40$

13. $x^2 + 120x + 2000$

14. $z^2 - 10z + 25$

In Examples 1-4, we factored trinomials of the type $x^2 + bx + c$. Now let's review factoring trinomials of the type $ax^2 + bx + c,\ a \neq 1$.

Example 5 Factor: $3x^2 + 10x - 8$.

Factor the first term. The only possibility is $3x \cdot x$.	$3x^2 + 10x - 8 = (3x \quad)(x \quad)$
Factor the last term, -8, which is negative.	The possibilities are $1(-8),\ (-1)8,\ (2)(-4)$, and $(-2)(4)$. Each can be written in either order.
Look for combinations of factors such that the sum of the outside and the inside products is the middle term $10x$.	There are 8 combinations of factors. $(3x+1)(x-8) = 3x^2 - 23x - 8$ $(3x-8)(x+1) = 3x^2 - 5x - 8$ $(3x-1)(x+8) = 3x^2 + 23x - 8$ $(3x+8)(x-1) = 3x^2 + 5x - 8$ $(3x+2)(x-4) = 3x^2 - 10x - 8$ $(3x-4)(x+2) = 3x^2 + 2x - 8$ $(3x-2)(x+4) = 3x^2 + 10x - 8 \quad \leftarrow$ $(3x+4)(x-2) = 3x^2 - 2x - 8$
Correct factorization.	$3x^2 + 10x - 8 = (3x-2)(x+4)$.

Example 6 Factor $21t^2 - 32t + 12$.

Factor the first term, $21t^2$. There are two possibilities: $21t \cdot t$ or $7t \cdot 3t$.	$21t^2 - 32t + 12 = (21t \quad)(t \quad)$ or $21t^2 - 32t + 12 = (7t \quad)(3t \quad)$
Factor the last term, 12, which is positive.	The possibilities are $12 \cdot 1$, $-12(-1)$, $6 \cdot 2$, $-6(-2)$, $4 \cdot 3$, and $-4(-3)$.
Since the middle term, $-32t$, is negative, consider only three of the six possibilities.	$-12(-1)$, $-6(-2)$, and $-4(-3)$. Each can be written in either order.
Begin by using $(7t \quad)(3t \quad)$. If a correct factorization is not found, then consider $(21t \quad)(t \quad)$. Look for combinations of factors such that the sum of the outside and the inside products is the middle term, $-32t$.	$(7t - 12)(3t - 1) = 21t^2 - 43t + 12$ $(7t - 1)(3t - 12) = 21t^2 - 87t + 12$ $(7t - 6)(3t - 2) = 21t^2 - 32t + 12$ $\leftarrow$ $(7t - 2)(3t - 6) = 21t^2 - 48t + 12$ $(7t - 4)(3t - 3) = 21t^2 - 33t + 12$ $(7t - 3)(3t - 4) = 21t^2 - 37t + 12$
Correct factorization.	$21t^2 - 32t + 12 = (7t - 6)(3t - 2)$.

Exercises Factor.

15. $6x^2 + 17x - 14$

16. $4w^2 - 4w - 15$

17. $30x^2 + 7x - 2$

18. $24y^2 - 29y - 4$

19. $20x^2 + 44x - 15$ 　　　　　　　　　20. $40c^2 + 31c + 6$

When the three terms of a trinomial have a common factor, we begin by factoring it out.

Example 7 Factor: $z^3 - 14z^2 + 33z$.

Factor out the common factor z.

$$z^3 - 14z^2 + 33z = z\left(z^2 - 14z + 33\right)$$

Now factor the trinomial $z^2 - 14z + 33$. The first term is z^2, so the first term of each binomial is z.

$$z^2 - 14z + 33 = (z \quad)(z \quad)$$

Find two numbers whose product is 33 and whose sum is -14.

Pairs of factors	Sum of factors
1, 33	34
$-1, -33$	-34
3, 11	14
$-3, -11$	-14 ←

The numbers we need are -3 and -11.

$$z^2 - 14z + 33 = (z - 3)(z - 11)$$

Don't forget the common factor, z.

$$z^3 - 14z^2 + 33z = z(z - 3)(z - 11)$$

Example 8 Factor: $3x^4 + 18x^3 - 21x^2$.

Factor out the common factor $3x^2$.	$3x^4 + 18x^3 - 21x^2 = 3x^2\left(x^2 + 6x - 7\right)$
Now factor the trinomial $x^2 + 6x - 7$. The first term is x^2, so the first term in each binomial is x.	$x^2 + 6x - 7 = (x \quad)(x \quad)$
Find two numbers whose product is -7 and whose sum is 6. Those numbers are -1 and 7.	$x^2 + 6x - 7 = (x - 1)(x + 7)$
Don't forget the common factor, $3x^2$.	$3x^4 + 18x^3 - 21x^2 = 3x^2(x - 1)(x + 7)$

Exercises Factor.

21. $x^3 - 5x^2 + 6x$

22. $b^3 - 4b^2 - 32b$

23. $2y^4 + 26y^3 - 60y^2$

24. $5q^3 + 65q^2 + 180q$

Integrated Review 5: FACTORING TRINOMIAL SQUARES AND DIFFERENCES OF SQUARES

Trinomial Squares

When the factorization of a trinomial is the square of a binomial, the trinomial is called a **trinomial square**, or perfect-square trinomial.

Examples:

$$x^2 + 14x + 49 = (x+7)(x+7) = (x+7)^2, \text{ and}$$

$$z^2 - 20z + 100 = (z-10)(z-10) = (z-10)^2$$

How to recognize a **trinomial square**, $A^2 + 2AB + B^2$ or $A^2 - 2AB + B^2$:

a) The two expressions A^2 and B^2 must be squares.

b) There must be no minus sign before either A^2 or B^2.

c) Multiplying A and B (expressions whose squares are A^2 and B^2) and doubling the result gives either the *remaining term* or its *opposite*.

Factoring Trinomial Squares

$$A^2 + 2AB + B^2 = (A+B)^2$$

$$A^2 - 2AB + B^2 = (A-B)^2$$

Example 1 Factor: $y^2 + 8y + 16$.

y^2 and 16 are squares: y^2 and 4^2.   ($A = y$, $B = 4$)

No minus sign before either y^2 or 16.

The remaining term $8y = 2 \cdot y \cdot 4$.
$2 \cdot A \cdot B$

$y^2 + 8y + 16$ is a trinomial square.

$y^2 + 8y + 16 = (y+4)^2$

Example 2 Factor: $25x^2 - 30x + 9$.

$25x^2$ and 9 are squares: $(5x)^2$ and 3^2.   ($A = 5x$, $B = 3$)

No minus sign before either $25x^2$ or 9.

The remaining term $-30x = -(2 \cdot 5x \cdot 3)$
$= -(2 \cdot A \cdot B)$ Think: The opposite of $2 \cdot A \cdot B$.

$25x^2 - 30x + 9$ is a trinomial square.

$25x^2 - 30x + 9 = (5x-3)^2$.

Differences of Squares

The following are differences of squares:

$$x^2 - 81, \ 16 - 9y^2, \ \text{and} \ 100w^2 - 1.$$

Before factoring a difference of squares, let's review finding the product of the sum and the difference of the same two terms.

$$(x+9)(x-9) = x^2 - 9x + 9x - 81 = x^2 - 81$$

We see that two terms $-9x$ and $9x$ are opposites. They add to 0 and "drop out." Looking at this product in reverse, we see that $x^2 - 81 = (x+9)(x-9)$.

Factoring A Difference or Squares

$$A^2 - B^2 = (A+B)(A-B)$$

Example 3 Factor: $x^2 - 36$.

$$x^2 - 36 = x^2 - 6^2$$
$$= (x+6)(x-6)$$

Example 4 Factor: $121c^2 - 4$.

$$121c^2 - 4 = (11c)^2 - 2^2$$
$$= (11c+2)(11c-2)$$

Check Your Understanding

Classify each of the following as a trinomial square, a difference of squares, or neither of these.

1. $100 - 9w^2$

2. $x^2 - 2x + 4$

3. $y^2 - 6y + 9$

4. $36x^2 - 1$

5. $4x^2 + 49$

6. $16x^2 + 40x + 25$

Exercises Factor.

1. $64x^2 - 1$

2. $x^2 - 12x + 36$

3. $81y^2 - 16$

4. $w^2 - 4w + 4$

5. $49x^2 + 42x + 9$

6. $25a^2 - 36$

7. $49 - z^2$

8. $x^2 - 225$

9. $y^2 + 16y + 64$

10. $4x^2 - 20x + 25$

11. $5x^2 - 45$

12. $3x^2 + 24x + 48$

Integrated Review 6: THE PRINCIPLE OF ZERO PRODUCTS

The principle of zero products gives us a method for solving polynomial equations.

The Principle of Zero Products

For any real numbers a and b:

If $ab = 0$, then $a = 0$ or $b = 0$ (or both).

If $a = 0$ or $b = 0$, then $ab = 0$.

Example 1 Solve: $x^2 - x - 6 = 0$.

Factor.	$(x-3)(x+2) = 0$
Set each factor equal to 0.	$x - 3 = 0$ *or* $x + 2 = 0$
Solve separately.	$x = 3$ *or* $x = -2$

Check.

$$\begin{array}{c|c} x^2 - x - 6 = 0 & \\ \hline 3^2 - 3 - 6 & 0 \\ 9 - 3 - 6 & \\ 0 & 0 \quad \text{True} \end{array} \qquad \begin{array}{c|c} x^2 - x - 6 = 0 & \\ \hline (-2)^2 - (-2) - 6 & 0 \\ 4 + 2 - 6 & \\ 0 & 0 \quad \text{True} \end{array}$$

The numbers 3 and −2 are both solutions.

Note that we must have 0 on one side of the equation in order to use the principle of zero products.

Example 2 Solve: $7y + 3y^2 = -2$.

Get 0 on one side and write in descending order.	$3y^2 + 7y + 2 = 0$
Factor.	$(3y+1)(y+2) = 0$
Use the principle of zero products.	$3y + 1 = 0$ *or* $y + 2 = 0$
Solve separately.	$3y = -1$ *or* $y = -2$
	$y = -\dfrac{1}{3}$ *or* $y = -2$

Check.

$$\begin{array}{c|c} 7y + 3y^2 = -2 & \\ \hline 7\left(-\dfrac{1}{3}\right) + 3\left(-\dfrac{1}{3}\right)^2 & -2 \\ 7\left(-\dfrac{1}{3}\right) + 3\left(\dfrac{1}{9}\right) & \\ -\dfrac{7}{3} + \dfrac{1}{3} & \\ -\dfrac{6}{3} & \\ -2 & -2 \quad \text{True} \end{array} \qquad \begin{array}{c|c} 7y + 3y^2 = -2 & \\ \hline 7(-2) + 3(-2)^2 & -2 \\ 7(-2) + 3(4) & \\ -14 + 12 & \\ -2 & -2 \quad \text{True} \end{array}$$

The solutions are $-\dfrac{1}{3}$ and -2.

Example 3 Solve: $5b^2 = 10b$.

Get 0 on one side.	$5b^2 - 10b = 0$
Factor.	$5b(b-2) = 0$
Use the principle of zero products.	$5b = 0$ *or* $b-2 = 0$
Solve separately.	$b = 0$ *or* $b = 2$
	The solutions are 0 and 2.

Example 4 Solve: $6x - x^2 = 9$.

Get 0 on one side and the leading coefficient on the other side positive.	$0 = x^2 - 6x + 9$
Factor.	$0 = (x-3)(x-3)$
Use the principle of zero products.	$x-3 = 0$ *or* $x-3 = 0$
Solve separately.	$x = 3$ *or* $x = 3$
	There is only one solution, 3.

Example 5 Solve: $3x^3 - 9x^2 = 30x$.

Get 0 on one side.	$3x^3 - 9x^2 - 30x = 0$
Factor out a common factor.	$3x(x^2 - 3x - 10) = 0$
Factor the trinomial.	$3x(x+2)(x-5) = 0$
Use the principle of zero products.	$3x = 0$ *or* $x+2 = 0$ *or* $x-5 = 0$
Solve separately.	$x = 0$ *or* $x = -2$ *or* $x = 5$
	The solutions are 0, −2, and 5.

Check Your Understanding

Determine whether each statement is true or false.

1. If $(s-6)(s+8) = 0$, then both $s-6$ and $s+8$ must equal 0.

2. If $(x+12)(x+9) = 0$, then $x+12 = 0$ or $x+9 = 0$.

3. If $(y+7)(y-3) = 21$, then $y+7 = 21$ or $y-3 = 21$.

Exercises

Solve using the principle of zero products. In Exercises 1 and 2, fill-in the blanks in key steps.

1.
$$x^2 - 7x = -10$$

$$x^2 - 7x + \boxed{} = 0$$

$$\left(x - \boxed{}\right)(x - 2) = 0$$

$$x - 5 = \boxed{} \quad or \quad x - 2 = \boxed{}$$

$$x = 5 \quad or \quad x = \boxed{}$$

The solutions are $\boxed{}$ and $\boxed{}$.

2.
$$15z^2 = -3z$$

$$15z^2 + 3z = 0$$

$$\boxed{}(5z + 1) = 0$$

$$3z = 0 \quad or \quad 5z + 1 = 0$$

$$z = \boxed{} \quad or \quad z = -\frac{1}{5}$$

The solutions are $\boxed{}$ and $\boxed{}$.

3. $y^2 + 2y = 63$

4. $18x^2 = 9x$

5. $9x + x^2 + 20 = 0$

6. $x^2 + 20x + 100 = 0$

7. $32 + 4x - x^2 = 0$

8. $11x + 4x^2 = -6$

9. $10 - r - 21r^2 = 0$

10. $2x^3 - 2x^2 = 12x$

Notes:

Integrated Review 7: INTERVAL NOTATION

The graph of an inequality is a drawing that represents its solutions. The following graph illustrates an inequality whose solution set is all real numbers less than 5.

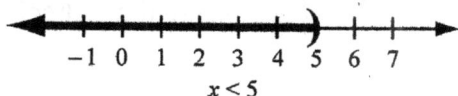

$$x < 5$$

The solution set can be written in set-builder notation, $\{x \mid x < 5\}$. This is read "The set of all x such that x is less than 5." The solution set can be also written in interval notation, $(-\infty, 5)$.

Interval notation uses parentheses ( ) and brackets [ ]. In Examples 1-9, intervals of the types (a, b), $[a, b]$, $[a, b)$, $(a, b]$, (a, ∞), $[a, \infty)$, $(-\infty, b)$, $(-\infty, b]$, and $(-\infty, \infty)$ are illustrated. The points a and b are endpoints of the interval. A parenthesis indicates that the endpoint is not included in the graph. A bracket indicates that the endpoint is included in the graph. Some intervals extend without bound in one or both directions. We use the symbol ∞, read "infinity," and $-\infty$, read "negative infinity," to name these intervals.

Examples

	Graph	Set Notation	Interval Notation
1.		$\{x \mid -5 < x < -2\}$	$(-5, -2)$
2.		$\{x \mid -1 \leq x \leq 3\}$	$[-1, 3]$
3.		$\{x \mid 0 \leq x < 4\}$	$[0, 4)$
4.		$\{x \mid 2 < x \leq 5\}$	$(2, 5]$
5.		$\{x \mid x > 1\}$	$(1, \infty)$
6.		$\{x \mid x \geq -3\}$	$[-3, \infty)$
7.		$\{x \mid x < -4\}$	$(-\infty, -4)$

8. 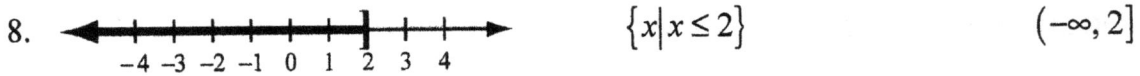 $\{x|x \le 2\}$ $(-\infty, 2]$

9. 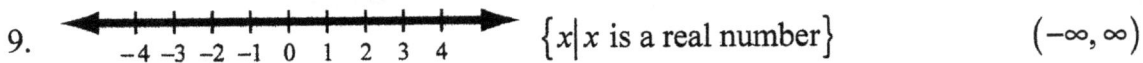 $\{x|x \text{ is a real number}\}$ $(-\infty, \infty)$

Check Your Understanding

For each set, consider the endpoints of the interval and choose from choices A – D, the format for interval notation.

A. (,)   B. [,]   C. [,)   D. (,]

1. $\{q|0 < q \le 9\}$

2. $\{x|-4 \le x < 1\}$

3. $\left\{y|y \ge \dfrac{3}{4}\right\}$

4. $\{t|0.7 < t < 2.5\}$

5. $\{x|-10 \le x \le 13\}$

6. $\{p|p < -2\}$

Exercises

Write interval notation for the given set or graph.

1.

2. $\{c|-4.1 \le c < 5.6\}$

3. $\{x|-8 < x \le 21\}$

4.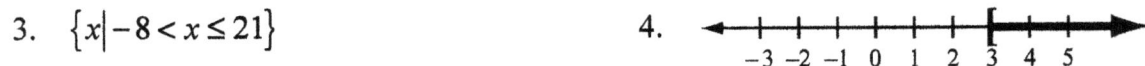

5. $\{w|w > 0\}$

6. $\{z|z \text{ is a real number}\}$

7.

8.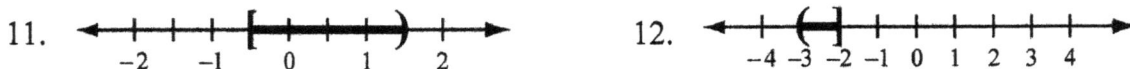

9. $\left\{y|y < \dfrac{2}{5}\right\}$

10. $\{d|d \ge 17\}$

11.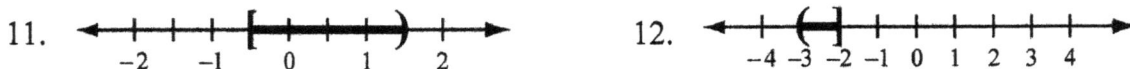

12.

Given a solution set expressed in interval notation, graph it on the number line.

13. $[3, 6)$

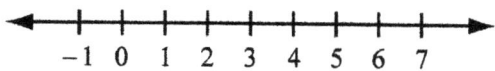

14. $[4, \infty)$

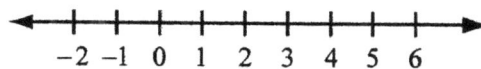

15. $(-1, \infty)$

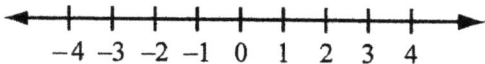

16. $(-1, 4]$

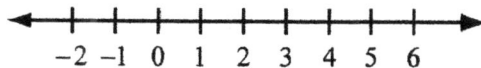

17. $[-5, -1]$

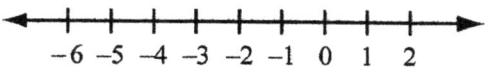

18. $(-\infty, \infty)$

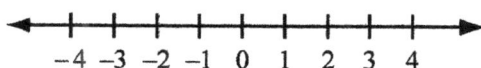

19. $(-\infty, -1.5)$

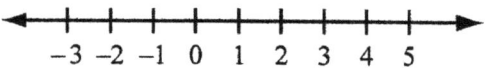

20. $(-\infty, 2]$

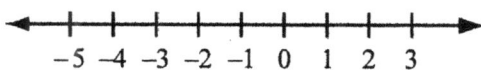

21. $[-12, 1)$

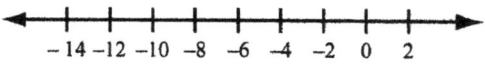

22. $(0, 20)$

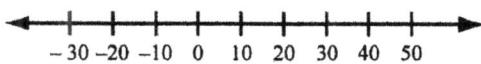

23. $(-15, \infty)$

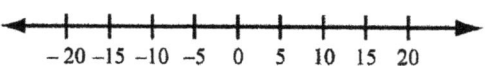

24. $(-3, 5]$

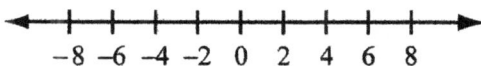

Notes:

Integrated Review 8: SOLVING LINEAR EQUATIONS

Equation Solving Principles

For any real numbers a, b, and c:

The Addition Principle: If $a = b$ is true, then $a + c = b + c$ is true.

The Multiplication Principle: If $a = b$ is true, then $ac = bc$ is true.

Using the Addition Principle

Example 1   Solve: $y - 3 = 20$.

Add 3 on both sides.	$y - 3 + 3 = 20 + 3$
Simplify.	$y + 0 = 23$
Identity property of 0: $a + 0 = a$	$y = 23$

Example 2   Solve: $8 + x = 5$.

Add -8, or subtract 8, on both sides.	$8 + x - 8 = 5 - 8$
Simplify.	$x + 0 = -3$
	$x = -3$

Using the Multiplication Principle

Example 3   Solve: $3z = -12$.

Multiply by $\dfrac{1}{3}$, or divide by 3, on both sides.	$\dfrac{3z}{3} = \dfrac{-12}{3}$
Simplify.	$1 \cdot z = -4$
Identity property of 1: $1 \cdot a = a$	$z = -4$

Example 4   Solve: $\dfrac{3}{2}x = 9$.

Multiply by $\dfrac{2}{3}$, the reciprocal of $\dfrac{3}{2}$.	$\dfrac{2}{3} \cdot \dfrac{3}{2} x = \dfrac{2}{3} \cdot 9$
Simplify.	$1 \cdot x = \dfrac{18}{3}$
	$x = 6$

Example 5 Solve: $-a = 1.4$.

$-a = -1 \cdot a$.	$-1 \cdot a = 1.4$
Divide by -1 on both sides.	$\dfrac{-1 \cdot a}{-1} = \dfrac{1.4}{-1}$
Simplify.	$a = -1.4$

Exercises Solve.

1. $w + 8 = 11$

2. $\dfrac{3}{5} - t = -\dfrac{2}{5}$

3. $-9z = 288$

4. $\dfrac{2}{7}y = 42$

5. $-x = 2.5$

6. $x - 4 = -10$

Using the Principles Together

Example 6   Solve: $5x - 6 = 29$.

Add 6 on both sides.	$5x - 6 + 6 = 29 + 6$
Simplify.	$5x = 35$
Divide by 5 on both sides.	$\dfrac{5x}{5} = \dfrac{35}{5}$
Simplify.	$x = 7$

Example 7   Solve: $3 + \dfrac{1}{2}y = -13$.

Subtract 3 on both sides.	$3 + \dfrac{1}{2}y - 3 = -13 - 3$
Simplify.	$\dfrac{1}{2}y = -16$
Multiply by 2 on both sides.	$2 \cdot \dfrac{1}{2}y = 2(-16)$
Simplify.	$y = -32$

Exercises Solve.

7. $9t + 4 = -104$

8. $4x - 7 = 81$

9. $11 = 2a - 15$

10. $5 - \dfrac{2}{3}y = -4$

Collecting Like Terms

Example 8   Solve: $6x + 5 - 7x = 10 - 4x + 3$.

Collect like terms on each side.	$-x + 5 = 13 - 4x$
Add $4x$.	$3x + 5 = 13$
Subtract 5.	$3x = 8$
Divide by 3.	$x = \dfrac{8}{3}$

Removing Parentheses

Example 9   Solve: $7(3x + 6) = 11 - (x + 2)$.

Remove parentheses using the distributive law.	$21x + 42 = 11 - x - 2$
Collect like terms on the right.	$21x + 42 = 9 - x$
Add x.	$22x + 42 = 9$
Subtract 42.	$22x = -33$
Divide by 22.	$x = \dfrac{-33}{22}$
Simplify.	$x = -\dfrac{3}{2}$

Exercises Solve.

11. $6y + 20 = 10 + 3y + y$

12. $5y - (2y - 10) = 25$

13. $80 = 10(3t + 2)$

14. $0.9x - 0.7x = 4.2$

15. $5 - 4a = a - 13$

16. $4(2x - 7) = 5 - (x + 3)$

Clearing Fractions

Example 10   Solve: $\dfrac{5}{6}y = \dfrac{1}{3}y - 7$

Multiply by 6, the least common multiple of 6 and 3.	$6\left(\dfrac{5}{6}y\right) = 6\left(\dfrac{1}{3}y - 7\right)$
Remove parentheses.	$5y = 2y - 42$
Subtract $2y$.	$3y = -42$
Divide by 3.	$y = -14$

Clearing Decimals

Example 11   Solve: $1.4 - 1.05x = 0.7x$

Greatest number of decimal places is 2. Multiply by 100.	$100(1.4 - 1.05x) = 100(0.7x)$
Remove parentheses.	$140 - 105x = 70x$
Add $105x$.	$140 = 175x$
Divide by 175.	$\dfrac{140}{175} = x$
Simplify.	$\dfrac{4}{5} = x$

Exercises Solve.

17. $5.4x + 2.04 = 1.5$

18. $\dfrac{5}{12}a - \dfrac{3}{8}a = \dfrac{1}{4}$

19. $\dfrac{3}{5}t - \dfrac{1}{10} = \dfrac{4}{15}t$

20. $6.7 + 0.001y = 9.82$

21. $2.1x + 45.2 = 3.2 - 8.4x$

22. $\dfrac{2}{7}x - \dfrac{1}{2}x = \dfrac{3}{4}x + 1$

Integrated Review 9: SOLVING LINEAR INEQUALITIES

Principles for Solving Inequalities

For any real numbers a, b, and c:

The Addition Principle for Inequalities:

If $a < b$ is true, then $a + c < b + c$ is true.

The Multiplication Principle for Inequalities:

a) If $a < b$ and $c > 0$ are true, then $ac < bc$ is true.
b) If $a < b$ and $c < 0$ are true, then $ac > bc$ is true.

    (When both sides of an inequality are multiplied by a negative number, the inequality sign must be reversed.)

Similar statements hold for $a \le b$.

Example 1 Solve: $x - 2 \ge -3$.

    Add 2. $\quad | \quad x - 2 + 2 \ge -3 + 2$

    Simplify. $\quad | \qquad x \ge -1$

Graph of solution:

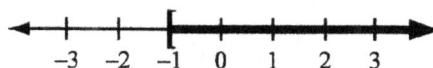

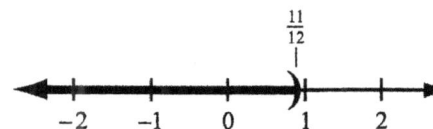

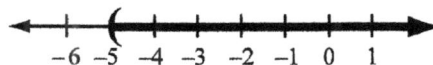

The solution set is $\{x \,|\, x \ge -1\}$, or $[-1, \infty)$.

Example 2 Solve: $y + \dfrac{1}{3} < \dfrac{5}{4}$.

    Subtract $\dfrac{1}{3}$. $\quad \left| \quad y + \dfrac{1}{3} - \dfrac{1}{3} < \dfrac{5}{4} - \dfrac{1}{3} \right.$

Multiply by 1 to obtain a common denominator.
$$y < \frac{5}{4} \cdot \frac{3}{3} - \frac{1}{3} \cdot \frac{4}{4}$$

$$y < \frac{15}{12} - \frac{4}{12}$$

$$y < \frac{11}{12}$$

Graph of solution:

The solution set is $\left\{ y \,|\, y < \dfrac{11}{12} \right\}$, or $\left(-\infty, \dfrac{11}{12} \right)$.

Example 3 Solve: $3x > -15$.

    Divide by 3. $\quad \left| \quad \dfrac{3x}{3} > \dfrac{-15}{3} \right.$

    Simplify. $\quad | \qquad x > -5$

Graph of solution:

The solution set is $\{x \,|\, x > -5\}$, or $(-5, \infty)$.

Example 4 Solve: $-4x \leq 8$.

Divide by -4. $\quad \dfrac{-4x}{-4} \geq \dfrac{8}{-4}$ $\leftarrow$ The symbol must be reversed.

Simplify. $\qquad x \geq -2$

The solution set is $\{x \mid x \geq -2\}$, or $[-2, \infty)$.

Graph of solution:

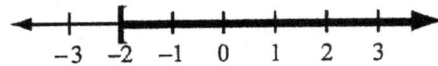

Example 5 Solve: $\quad -\dfrac{2}{3}a > -\dfrac{1}{10}$.

Multiply by $-\dfrac{3}{2}$. $\quad \left|\; -\dfrac{3}{2}\left(-\dfrac{2}{3}a\right) < -\dfrac{3}{2}\left(-\dfrac{1}{10}\right)\right.$ $\leftarrow$ The symbol must be reversed.

Simplify. $\qquad\qquad a < \dfrac{3}{20}$

The solution set is $\left\{a \mid a < \dfrac{3}{20}\right\}$, or $\left(-\infty, \dfrac{3}{20}\right)$.

Example 6 Solve: $10 - 6x > 3x - 8$.

Add 8. $\quad 18 - 6x > 3x$

Add $6x$. $\quad 18 > 9x$

Divide by 9. $\quad 2 > x$

The solution set is $\{x \mid 2 > x\}$, or $\{x \mid x < 2\}$, or $(-\infty, 2)$.

Example 7 Solve: $2(t - 5) - 8 \geq 3t - 7$.

Remove parentheses. $\quad 2t - 10 - 8 \geq 3t - 7$

Collect like terms. $\quad 2t - 18 \geq 3t - 7$

Subtract $3t$. $\quad -1t - 18 \geq -7$

Add 18. $\quad -t \geq 11$

Multiply by -1 and reverse the symbol. $\quad t \leq -11$

The solution set is $\{t \mid t \leq -11\}$, or $(-\infty, -11]$.

Exercises Solve.

1. $y + 8 \geq 5$

2. $x - \dfrac{4}{5} > \dfrac{3}{4}$

3. $2c < -12$

4. $-7q \leq -42$

5. $-\dfrac{5}{8}x > \dfrac{5}{12}$

6. $4z - 10 < z + 26$

7. $x - 1 \geq 1 - x$

8. $23 - 5x \geq 2x - 12$

9. $8t - 6 \leq 2(t + 9)$

10. $-3(x + 1) - 2 < 3 - x$

Integrated Review 10: THE PYTHAGOREAN THEOREM

A **right triangle** is a triangle with a 90° angle, as shown here. In a right triangle, the longest side is called the **hypotenuse**. It is the side opposite the right angle. The other two sides are called **legs**. We generally use the letters a and b for the lengths of the legs and c for the length of the hypotenuse. They are related as follows.

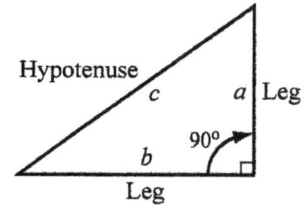

The Pythagorean Theorem

In any right triangle, if a and b are the lengths of the legs and c is the length of the hypotenuse, then

$$a^2 + b^2 = c^2, \text{ or}$$

$$\left(\text{Leg}\right)^2 + \left(\text{Other leg}\right)^2 = \left(\text{Hypotenuse}\right)^2.$$

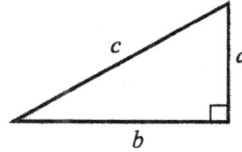

If we know the lengths of any two sides of a right triangle, we can use the Pythagorean theorem to determine the length of the third side.

Example 1 Find the length of the hypotenuse of this right triangle.

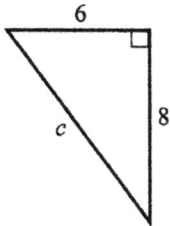

$$a^2 + b^2 = c^2$$
$$6^2 + 8^2 = c^2 \qquad \text{Substituting}$$
$$36 + 64 = c^2$$
$$100 = c^2$$

The solution of this equation is the square root of 100, which is 10.

$$c = \sqrt{100} = 10.$$

Example 2 Find the length b for the right triangle shown. Give an exact answer and an approximation to three decimal places.

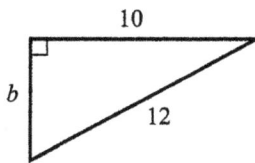

$$a^2 + b^2 = c^2$$
$$10^2 + b^2 = 12^2 \qquad \text{Substituting}$$
$$100 + b^2 = 144.$$
$$100 + b^2 - 100 = 144 - 100 \qquad \text{Subtracting 100 on both sides}$$
$$b^2 = 44$$

Exact answer: $b = \sqrt{44}$

Approximation: $b \approx 6.633.$ \qquad Using a calculator

Exercises Find the length of the third side of each right triangle. Give an exact answer and, when appropriate, an approximation to three decimal places.

1.

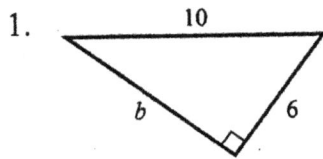

2.

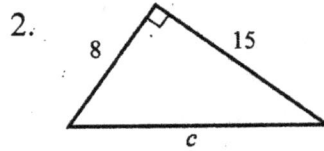

3.

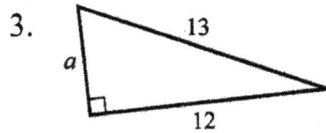

4.

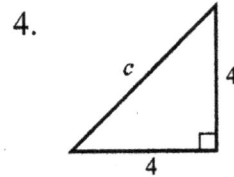

5.

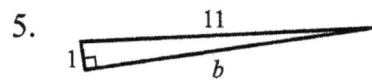

6.

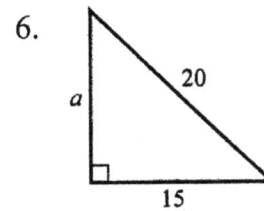

7.

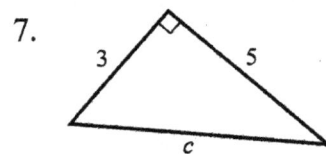

8.

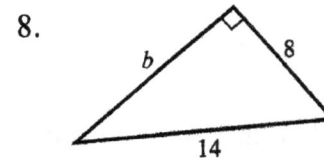

9.

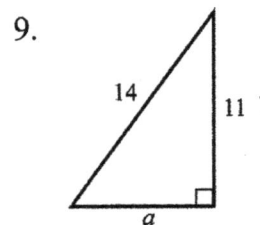

10.

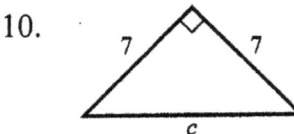

Integrated Review 11: SIMPLIFYING RATIONAL EXPRESSIONS

We simplify rational expressions using the identity property of 1 in reverse. That is, we "remove" factors that are equal to 1. We first factor the numerator and the denominator and then factor the rational expression, so that a factor is equal to 1.

Example 1   Simplify: $\dfrac{120}{320}$.

Factor the numerator and the denominator.	$= \dfrac{40 \cdot 3}{40 \cdot 8}$
Factor the rational expression.	$= \dfrac{40}{40} \cdot \dfrac{3}{8}$
Remove a factor of 1, $\dfrac{40}{40} = 1$.	$= 1 \cdot \dfrac{3}{8} = \dfrac{3}{8}$

Example 2   Simplify: $\dfrac{5x^2}{x}$.

Factor the numerator and the denominator.	$= \dfrac{5x \cdot x}{1 \cdot x}$
Factor the rational expression.	$= \dfrac{5x}{1} \cdot \dfrac{x}{x}$
Remove a factor of 1, $\dfrac{x}{x} = 1$.	$= 5x \cdot 1 = 5x$

Example 3   Simplify: $\dfrac{2x^2 + 4x}{6x^2 + 2x}$.

Factor the numerator and the denominator.	$= \dfrac{2x(x+2)}{2x(3x+1)}$
Factor the rational expression.	$= \dfrac{2x}{2x} \cdot \dfrac{x+2}{3x+1}$
Remove a factor of 1, $\dfrac{2x}{2x} = 1$.	$= 1 \cdot \dfrac{x+2}{3x+1} = \dfrac{x+2}{3x+1}$

Exercises Simplify.

1. $\dfrac{60}{75}$

2. $\dfrac{6w^3}{2w^2}$

3. $\dfrac{4x^2 + 16x}{36x^2 + 4x}$

4. $\dfrac{x^3 + 10x^2}{x^3 - 9x^2}$

5. $\dfrac{y}{9y^7}$

6. $\dfrac{12z^3 - 15z^2}{3z^3 + 45z^2}$

Example 4 Simplify: $\dfrac{x^2+2x}{3x+6}$.

Factor the numerator and the denominator. $= \dfrac{x(x+2)}{3(x+2)}$

Factor the rational expression. $= \dfrac{x}{3} \cdot \dfrac{x+2}{x+2}$

Remove a factor of 1, $\dfrac{x+2}{x+2} = 1$. $= \dfrac{x}{3} \cdot 1 = \dfrac{x}{3}$

Example 5 Simplify: $\dfrac{q^2-5q-66}{q^2+15q+54}$.

Factor the numerator and the denominator. $= \dfrac{(q+6)(q-11)}{(q+6)(q+9)}$

Factor the rational expression. $= \dfrac{q+6}{q+6} \cdot \dfrac{q-11}{q+9}$

Remove a factor of 1, $\dfrac{q+6}{q+6} = 1$. $= \dfrac{q-11}{q+9}$

Example 6 Simplify: $\dfrac{x-3}{3-x}$.

Rewrite $3-x$ as $-(x-3)$. $= \dfrac{x-3}{-(x-3)}$

Factor the numerator and the denominator. $= \dfrac{1(x-3)}{-1(x-3)}$

Simplify. $= -1$

Exercises Simplify.

7. $\dfrac{y^2-49}{y^2+11y+28}$

8. $\dfrac{6x^2-6x}{3x^3-3x^2}$

9. $\dfrac{a^2-a-6}{a^2-11a+24}$

10. $\dfrac{8y+24}{y^2+3y}$

11. $\dfrac{2-t}{t-2}$

12. $\dfrac{3x^2+11x-4}{2x^2+9x+4}$

Integrated Review 12: ADDING AND SUBTRACTING POLYNOMIALS

Adding Polynomials

When two terms have the same variable(s) raised to the same power(s), they are called **like terms**, or **similar terms**, and they can be "collected," or "combined," using the distributive laws. The sum of two polynomials can be found by writing a plus sign between them and then collecting like terms to simplify the expression.

Example 1   Add: $\left(3x^3 + 4x^2 - 7x - 2\right) + \left(-7x^3 - 2x^2 + 3x + 4\right)$.

Collect like terms. $= (3-7)x^3 + (4-2)x^2 + (-7+3)x + (-2+4)$

Simplify. $= -4x^3 + 2x^2 - 4x + 2$

Example 2   Add: $\left(8t^4 - t + 3t^3 - 8\right) + \left(5t - 3t^2 - t^3 + 2t^4\right)$.

Collect like terms. $= (8+2)t^4 + (3-1)t^3 - 3t^2 + (-1+5)t - 8$

Simplify. $= 6t^4 + 2t^3 - 3t^2 + 4t - 8$

Subtracting Polynomials

To subtract a real number, we add its *opposite*.

$$7 - (-3) = 7 + 3 = 10 \qquad\qquad -9 - 4 = -9 + (-4) = -13$$

To subtract a polynomial, we also add its *opposite*. We can find the opposite of a polynomial by changing the sign of *each* term. The opposite of $5x^2 - 4x - 3$ is $-5x^2 + 4x + 3$.

$-\left(5x^2 - 4x - 3\right)$ is equivalent to $-5x^2 + 4x + 3$.

Example 3 Subtract: $\left(2x^2 - 5x + 3\right) - \left(x^2 - 9x + 10\right)$.

Add the opposite. $= \left(2x^2 - 5x + 3\right) + \left(-x^2 + 9x - 10\right)$

Collect like terms. $= (2-1)x^2 + (-5+9)x + (3-10)$

Simplify. $= x^2 + 4x - 7$

Example 4 Subtract: $\left(y^3 - 1\right) - \left(y^3 + 2y + 6\right)$.

Add the opposite. $= \left(y^3 - 1\right) + \left(-y^3 - 2y - 6\right)$

Collect like terms. $= \left[1 + (-1)\right]y^3 - 2y + \left[-1 + (-6)\right]$

Simplify. $= -2y - 7$

Example 5 Subtract: $\left(-9x^3 + x^2 - 5x\right) - \left(-3x^4 + 2x^3 - 7x - 15\right)$.

Add the opposite. $= \left(-9x^3 + x^2 - 5x\right) + \left(3x^4 - 2x^3 + 7x + 15\right)$

Collect like terms. $= 3x^4 + \left[-9 + (-2)\right]x^3 + x^2 + (-5+7)x + 15$

Simplify. $= 3x^4 - 11x^3 + x^2 + 2x + 15$

Check Your Understanding

For each subtraction, choose an equivalent expression from choices a) – l).

a) $\left(4y^3 + 2y^2\right) + \left(y^3 + 8y - 4\right)$

b) $(w-3) + (-2w+11)$

c) $\left(p^5 - p^3 - p\right) + (3p+5)$

d) $\left(7x^2 + x + 4\right) + \left(-5x^2 + 2x - 1\right)$

e) $\left(7x^2 - x - 4\right) + \left(5x^2 - 2x - 1\right)$

f) $(w-3) + (-2w-11)$

g) $(w+3) + (2w-11)$

h) $\left(4y^3 - 2y^2\right) - \left(-y^3 + 8y - 4\right)$

i) $\left(-4y^3 + 2y^2\right) + \left(-y^3 + 8y - 4\right)$

j) $\left(p^5 - p^3 - p\right) + (-3p+5)$

k) $\left(p^5 - p^3 - p\right) + (3p-5)$

l) $\left(7x^2 - x - 4\right) + \left(5x^2 - 2x + 1\right)$

1. $(w-3) - (2w+11)$

2. $\left(7x^2 - x - 4\right) - \left(-5x^2 + 2x - 1\right)$

3. $\left(-4y^3 + 2y^2\right) - \left(y^3 - 8y + 4\right)$

4. $\left(p^5 - p^3 - p\right) - (-3p-5)$

Exercises Perform the indicated operation and simplify.

1. $\left(t^2 - 6t + 5\right) + \left(-3t^2 - 8\right)$

2. $(5x-2) - (x-7)$

3. $\left(2x^2 - 3x - 4\right) - \left(-7x^2 + 12\right)$

4. $\left(11y^2 + 6y - 3\right) + \left(9y^2 - 2y + 9\right)$

5. $\left(8b^3 - 11b\right) - \left(-2b^3 + b^2 + 3\right)$

6. $\left(3x^4 - 4x^2 + x - 3\right) - \left(5x^3 + 7x^2 + 1\right)$

7. $\left(15x^2 + 12x + 1\right) - \left(-x^2 + 3x + 2\right)$

8. $\left(7a^3 - 5a + 10\right) + \left(2a^4 - 9a^2 + a - 3\right)$

9. $\left(-10x^4 + x^3 - 2\right) + \left(-5x^3 + 2x - 6\right)$

10. $\left(w^5 - w^4 + w^2 - 1\right) - \left(-w^4 + 5w^3 - w^2 - 5\right)$

Integrated Review 13: SIMPLIFYING COMPLEX RATIONAL EXPRESSIONS

A **complex rational expression** is a rational expression that contains rational expressions within its numerator and/or its denominator. Here are some examples:

$$\frac{\dfrac{3}{10}-\dfrac{1}{6}}{\dfrac{7}{8}}, \quad \frac{\dfrac{1}{t}+6}{\dfrac{1}{t-5}}, \quad \frac{y-\dfrac{6}{y}}{y+\dfrac{6}{y}}, \quad \frac{1+\dfrac{1}{x}}{1-\dfrac{1}{x^2}}, \quad \frac{\dfrac{1}{x+h}-\dfrac{1}{x}}{h}$$

Example 1 Simplify: $\dfrac{\dfrac{3}{10}-\dfrac{1}{6}}{\dfrac{7}{8}}$.

The LCD in the numerator is 30. We multiply by 1. Then subtract in the numerator.

$$=\frac{\dfrac{3}{10}\cdot\dfrac{3}{3}-\dfrac{1}{6}\cdot\dfrac{5}{5}}{\dfrac{7}{8}}=\frac{\dfrac{9}{30}-\dfrac{5}{30}}{\dfrac{7}{8}}=\frac{\dfrac{4}{30}}{\dfrac{7}{8}}$$

Multiply by the reciprocal of the denominator.

$$=\frac{4}{30}\cdot\frac{8}{7}=\frac{4\cdot 8}{30\cdot 7}=\frac{32}{210}$$

Factor.

$$=\frac{2\cdot 16}{2\cdot 105}=\frac{2}{2}\cdot\frac{16}{105}$$

Remove a factor of 1.

$$=\frac{16}{105}$$

Example 2 Simplify: $\dfrac{\dfrac{1}{t}+6}{\dfrac{1}{t}-5}$.

The LCD in the numerator is t. The LCD in the denominator is t. We multiply by 1.

$$=\frac{\dfrac{1}{t}+6\cdot\dfrac{t}{t}}{\dfrac{1}{t}-5\cdot\dfrac{t}{t}}=\frac{\dfrac{1}{t}+\dfrac{6t}{t}}{\dfrac{1}{t}-\dfrac{5t}{t}}$$

Add in the numerator and subtract in the denominator.

$$=\frac{\dfrac{1+6t}{t}}{\dfrac{1-5t}{t}}$$

Multiply by the reciprocal of the denominator.

$$=\frac{1+6t}{t}\cdot\frac{t}{1-5t}=\frac{t(1+6t)}{t(1-5t)}$$

Factor.

$$=\frac{t}{t}\cdot\frac{1+6t}{1-5t}$$

Remove a factor of 1.

$$=\frac{1+6t}{1-5t}$$

Exercises Simplify.

1. $\dfrac{\dfrac{y}{2}}{7} =$

2. $\dfrac{\dfrac{3}{x}}{\dfrac{2}{x}} =$

3. $\dfrac{\dfrac{3}{8} - \dfrac{1}{12}}{\dfrac{5}{6}} =$

4. $\dfrac{\dfrac{2}{3} + \dfrac{4}{5}}{\dfrac{3}{4} - \dfrac{1}{2}} =$

5. $\dfrac{\dfrac{1}{x} - 9}{\dfrac{1}{x} + 2} =$

6. $\dfrac{3 - \dfrac{5}{z}}{4 + \dfrac{3}{z}} =$

Example 3   Simplify: $\dfrac{y - \dfrac{6}{y}}{y + \dfrac{6}{y}}$.

The LCD in the numerator is y. The LCD in the denominator is y. We multiply by 1.	$= \dfrac{\dfrac{y}{1} \cdot \dfrac{y}{y} - \dfrac{6}{y}}{\dfrac{y}{1} \cdot \dfrac{y}{y} + \dfrac{6}{y}} = \dfrac{\dfrac{y^2}{y} - \dfrac{6}{y}}{\dfrac{y^2}{y} + \dfrac{6}{y}}$
Subtract in the numerator and add in the denominator.	$= \dfrac{\dfrac{y^2 - 6}{y}}{\dfrac{y^2 + 6}{y}}$
Multiply by the reciprocal of the denominator.	$= \dfrac{y^2 - 6}{y} \cdot \dfrac{y}{y^2 + 6} = \dfrac{y(y^2 - 6)}{y(y^2 + 6)}$
Factor and remove a factor of 1.	$= \dfrac{y}{y} \cdot \dfrac{y^2 - 6}{y^2 + 6} = \dfrac{y^2 - 6}{y^2 + 6}$

Example 4   Simplify: $\dfrac{1+\dfrac{1}{x}}{1-\dfrac{1}{x^2}}$.

The LCD in the numerator is x.
The LCD in the denominator is x^2. We multiply by 1.

$$= \dfrac{1\cdot\dfrac{x}{x}+\dfrac{1}{x}}{1\cdot\dfrac{x^2}{x^2}-\dfrac{1}{x^2}} = \dfrac{\dfrac{x}{x}+\dfrac{1}{x}}{\dfrac{x^2}{x^2}-\dfrac{1}{x^2}}$$

Add in the numerator and subtract in the denominator.

$$= \dfrac{\dfrac{x+1}{x}}{\dfrac{x^2-1}{x^2}}$$

Multiply by the reciprocal of the denominator.

$$= \dfrac{x+1}{x}\cdot\dfrac{x^2}{x^2-1} = \dfrac{x^2(x+1)}{x(x^2-1)}$$

Factor and remove a factor of 1.

$$= \dfrac{x\cdot x(x+1)}{x(x+1)(x-1)} = \dfrac{x(x+1)}{x(x+1)}\cdot\dfrac{x}{x-1} = \dfrac{x}{x-1}$$

Exercises

7. $\dfrac{p-\dfrac{5}{p}}{p+\dfrac{5}{p}} =$

8. $\dfrac{\dfrac{1}{a}+7a}{\dfrac{1}{a}+a} =$

9. $\dfrac{4-\dfrac{1}{x^2}}{2+\dfrac{1}{x}} =$

10. $\dfrac{5-\dfrac{1}{y}}{25-\dfrac{1}{y^2}} =$

11. $\dfrac{\dfrac{1}{a}+\dfrac{1}{b}}{\dfrac{1}{a^2}-\dfrac{1}{b^2}} =$

Example 5   Simplify: $\dfrac{\dfrac{1}{x+h}-\dfrac{1}{x}}{h}$.

The LCD in the numerator is $x(x+h)$. We multiply by 1.

$$=\dfrac{\dfrac{1}{x+h}\cdot\dfrac{x}{x}-\dfrac{1}{x}\cdot\dfrac{x+h}{x+h}}{h}=\dfrac{\dfrac{x}{x(x+h)}-\dfrac{x+h}{x(x+h)}}{h}$$

Subtract in the numerator; $-(x+h)=-x-h$. It helps to rewrite the h in the denominator as $\dfrac{h}{1}$.

$$=\dfrac{\dfrac{x-(x+h)}{x(x+h)}}{\dfrac{h}{1}}=\dfrac{\dfrac{x-x-h}{x(x+h)}}{\dfrac{h}{1}}=\dfrac{\dfrac{-h}{x(x+h)}}{\dfrac{h}{1}}$$

Multiply by the reciprocal of the denominator.

$$=\dfrac{-h}{x(x+h)}\cdot\dfrac{1}{h}=\dfrac{-h}{x(x+h)h}$$

Factor and remove a factor of 1.

$$=\dfrac{-1}{x(x+h)}\cdot\dfrac{h}{h}=\dfrac{-1}{x(x+h)},\ \text{or}\ -\dfrac{1}{x(x+h)}$$

Exercises Simplify.

12. $\dfrac{\dfrac{1}{z+h}-\dfrac{1}{z}}{h}=$

13. $\dfrac{\dfrac{2}{x+h}-\dfrac{2}{x}}{h}=$

Integrated Review 14: MULTIPLYING BINOMIALS

The Product of Two Binomials Using the FOIL Method

Consider $(x+7)(x+4)$, the product of two binomials. We multiply each term of $(x+7)$ by each term of $(x+4)$:

$$(x+7)(x+4) = x \cdot x + x \cdot 4 + 7 \cdot x + 7 \cdot 4.$$

The multiplication illustrates a pattern that occurs whenever two binomials are multiplied:

$$\underbrace{\text{First}}_{} \quad \underbrace{\text{Outside}}_{} \quad \underbrace{\text{Inside}}_{} \quad \underbrace{\text{Last}}_{}$$
terms   terms   terms   terms

$$(x+7)(x+4) = x \cdot x \ + \ x \cdot 4 \ + \ 7 \cdot x \ + \ 7 \cdot 4 = x^2 + 11x + 28.$$

This special method of multiplying is called the **FOIL method**. Keep in mind that this method is based on the distributive law.

The Foil Method

To multiply two binomials, $A + B$ and $C + D$, multiply the **F**irst terms AC, the **O**utside terms AD, the **I**nside terms BC, and then the **L**ast terms BD. Then collect like terms, if possible.

$$(A+B)(C+D) = AC + AD + BC + BD$$

1. Multiply **F**irst terms: AC.
2. Multiply **O**utside terms: AD.
3. Multiply **I**nside terms: BC.
4. Multiply **L**ast terms: BD.

$$\downarrow$$
FOIL

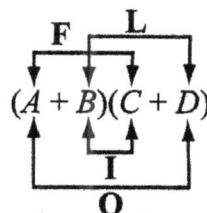

Example 1   Multiply: $(x+8)(x-5)$.

Use the FOIL method.	$\begin{aligned} &\quad\text{F}\qquad\text{O}\qquad\text{I}\qquad\text{L} \\ &= x \cdot x + x(-5) + 8 \cdot x + 8(-5) \end{aligned}$
Simplify.	$= x^2 - 5x + 8x - 40$
Collect like terms.	$= x^2 + 3x - 40$

Example 2   Multiply: $(3x-2)(6x-7)$.

Use the FOIL method.	$\begin{aligned} &\quad\text{F}\qquad\text{O}\qquad\quad\text{I}\qquad\quad\text{L} \\ &= 3x \cdot 6x + 3x(-7) + (-2)(6x) + (-2)(-7) \end{aligned}$
Simplify.	$= 18x^2 - 21x - 12x + 14$
Collect like terms.	$= 18x^2 - 33x + 14$

Exercises Multiply.

1. $(y-2)(y-3)$

2. $(2b-5)(8b+1)$

3. $(w+5)(w-2)$

4. $(z-9)(z+1)$

5. $(4x+3)(5x-6)$

6. $(3q+5)(q+20)$

7. $(x-4)(x+13)$

8. $(10b-7)(3b+4)$

9. $(5w-1)(5w+8)$

10. $(a+6)(a+11)$

Squares of Binomials

We use the FOIL method to develop special products for the square of a binomial:

$$(x+5)^2 = (x+5)(x+5) = x^2 + 5x + 5x + 5^2 = x^2 + 10x + 25,$$
$$(y-5)^2 = (y-5)(y-5) = y^2 - 5y - 5y + (-5)^2 = y^2 - 10y + 25.$$

Square of a Binomial

The **square of a binomial** is the square of the first term, plus twice the product of the two terms, plus the square of the last term.

$$(A+B)^2 = A^2 + 2AB + B^2;$$
$$(A-B)^2 = A^2 - 2AB + B^2$$

Example 3   Multiply: $(y-7)^2$.

Use $(A-B)^2 = A^2 - 2AB + B^2$. $= y^2 - 2 \cdot y \cdot 7 + 7^2$

                     Simplify. $= y^2 - 14y + 49$

Example 4   Multiply: $(3x+4)^2$.

Use $(A+B)^2 = A^2 + 2AB + B^2$. $\quad = (3x)^2 + 2 \cdot 3x \cdot 4 + 4^2$
$\qquad\qquad\qquad$ Simplify. $\quad = 9x^2 + 24x + 16$

Exercises Multiply.

11. $(x+6)^2$

12. $(2x-1)^2$

13. $(3a+2)^2$

14. $(t+10)^2$

15. $(z-12)^2$

16. $(4w+5)^2$

17. $(b-3)^2$

18. $(6y-3)^2$

Products of Sums and Differences

Another special case of a product of two binomials is the product of a sum and a difference. Note the following:

$$(x+9)(x-9) = x^2 - 9x + 9x + 9(-9) = x^2 - 81.$$

Product of a Sum and a Difference

The product of the sum and the difference of the same two terms is the square of the first term minus the square of the second term (the difference of their squares).

$$(A+B)(A-B) = A^2 - B^2 \quad \text{This is called a \textbf{difference of squares}.}$$

Example 5   Multiply: $(x+4)(x-4)$.

Use $(A+B)(A-B)=A^2-B^2$. $\quad = x^2 - 4^2$

Simplify. $\quad = x^2 - 16$

Example 6   Multiply: $(5z+8)(5z-8)$.

Use $(A+B)(A-B)=A^2-B^2$. $\quad = (5z)^2 - 8^2$

Simplify. $\quad = 25z^2 - 64$

Exercises Multiply.

19. $(y+10)(y-10)$

20. $(3c+4)(3c-4)$

21. $(t-6)(t+6)$

22. $(w+1)(w-1)$

23. $(2z+7)(2z-7)$

24. $(10y-5)(10y+5)$

Try to multiply polynomials mentally. When several types are mixed, first check to see what types of polynomials are to be multiplied. Then use the quickest method. Sometimes we might use more than one method to find a product. Remember that FOIL *always* works for multiplying binomials!

Exercises Multiply.

25. $(x-3)(x+10)$

26. $(4x+9)(4x-9)$

27. $(3x-1)(7x-4)$

28. $(y+20)(y+5)$

29. $(s-3)(s+3)$

30. $(a-8)^2$

31. $(5y+4)^2$

32. $(6a-2)(a+11)$

Integrated Review 15: SIMPLIFYING RADICAL EXPRESSIONS

For any nonnegative radicands A and B,

$$\sqrt{A} \cdot \sqrt{B} = \sqrt{A \cdot B}$$

The product of square roots is the square root of the product of the radicands. To factor radical expressions, we can use the product rule for radicals in reverse.

$$\sqrt{AB} = \sqrt{A}\sqrt{B}$$

When simplifying a square-root radical expression, if the radicand is not a perfect square, we determine whether it has perfect-square factors. If so, the radicand is then factored and the radical expression simplified using the preceding rule. A square-root radical expression is simplified when its radicand has no factors that are perfect squares.

Example 1   Simplify: $\sqrt{18}$.

Identify a perfect-square factor and factor the radicand.	$= \sqrt{9 \cdot 2}$
Factor into a product of radicals.	$= \sqrt{9} \cdot \sqrt{2}$
Simplify $\sqrt{9}$.	$= 3\sqrt{2}$

Exercises Simplify by factoring.

1. $\sqrt{12}$ 2. $\sqrt{75}$ 3. $\sqrt{80}$ 4. $\sqrt{72}$

5. $\sqrt{363}$ 6. $\sqrt{450}$ 7. $\sqrt{320}$ 8. $\sqrt{600}$

Example 2   Simplify: $\sqrt{48x}$.

Identify a perfect-square factor and factor the radicand.	$= \sqrt{16 \cdot 3 \cdot x}$
Factor into a product of radicals.	$= \sqrt{16} \cdot \sqrt{3x}$
Simplify.	$= 4\sqrt{3x}$

For our work here, we assume that expressions under radicals do no represent the square of a negative number. Thus, absolute-value signs are not necessary.

Example 3   Simplify: $\sqrt{20t^2}$.

Factor the radicand.	$= \sqrt{4 \cdot 5 \cdot t^2}$
Factor into a product of radicals.	$= \sqrt{4} \cdot \sqrt{t^2} \cdot \sqrt{5}$
Simplify.	$= 2t\sqrt{5}$

Example 4 Simplify: $\sqrt{x^2-8x+16}$.

Factor the radicand. $\quad = \sqrt{(x-4)^2}$

Simplify. $\quad = x-4$

Example 5 Simplify: $\sqrt{a^{10}}$.

Factor the radicand; note that $a^{10}=a^5\cdot a^5$. $\quad = \sqrt{\left(a^5\right)^2}$

Simplify. $\quad = a^5$

Example 6   Simplify: $\sqrt{24x^{17}}$.

Factor the radicand; note that $x^{17}=x^{16}\cdot x$. $\quad = \sqrt{4\cdot 6\cdot x^{16}\cdot x}$

Factor into a product of radicals. Note that $x^{16}=x^8\cdot x^8$. $\quad = \sqrt{4}\cdot\sqrt{\left(x^8\right)^2}\cdot\sqrt{6x}$

Simplify. $\quad = 2x^8\sqrt{6x}$

Exercises Simplify by factoring.

9. $\sqrt{25x}$

10. $\sqrt{y^{22}}$

11. $\sqrt{36x^2}$

12. $\sqrt{a^2+2a+1}$

13. $\sqrt{84t^4}$

14. $\sqrt{x^2-10x+25}$

15. $\sqrt{w^9}$

16. $\sqrt{32x^2}$

17. $\sqrt{x^2-24x+144}$

18. $\sqrt{3y^2+6y+3}$

19. $\sqrt{1400b^6}$

20. $\sqrt{81x^7}$

Integrated Review 16: MULTIPLYING RADICAL EXPRESSIONS

> **Product Rule for Radicals**
>
> For any nonnegative radicands A and B,
>
> $$\sqrt{A} \cdot \sqrt{B} = \sqrt{A \cdot B}.$$
>
> (The product of square roots is the square root of the product of the radicands.)

Example 1 Multiply and simplify: $\sqrt{2}\sqrt{14}$.

Multiply radicands.	$= \sqrt{2 \cdot 14}$
Factor.	$= \sqrt{2 \cdot 2 \cdot 7}$
Look for pairs of factors.	$= \sqrt{2 \cdot 2}\sqrt{7}$
Simplify.	$= 2\sqrt{7}$

Example 2 Multiply and simplify: $\sqrt{45}\sqrt{15}$.

Multiply radicands.	$= \sqrt{45 \cdot 15}$
Factor.	$= \sqrt{5 \cdot 3 \cdot 3 \cdot 3 \cdot 5}$
Look for pairs of factors.	$= \sqrt{5 \cdot 5}\sqrt{3 \cdot 3}\sqrt{3}$
Simplify.	$= 5 \cdot 3 \cdot \sqrt{3}$
	$= 15\sqrt{3}$

In this course, we will assume that no radicands are formed by raising negative quantities to even powers. Thus, absolute value signs are not necessary.

Example 3 Multiply and simplify: $\sqrt{3}\sqrt{2y+5}$.

Multiply radicands. We cannot factor further.	$= \sqrt{3(2y+5)}$
Simplify.	$= \sqrt{6y+15}$

Example 4 Multiply and simplify: $\sqrt{6t}\sqrt{30t}$

Multiply radicands.	$= \sqrt{6t \cdot 30t}$
Factor.	$= \sqrt{2 \cdot 3 \cdot t \cdot 2 \cdot 3 \cdot 5 \cdot t}$
Look for pairs of factors.	$= \sqrt{2 \cdot 2}\sqrt{3 \cdot 3}\sqrt{t \cdot t}\sqrt{5}$
	$= 2 \cdot 3 \cdot t\sqrt{5}$
Simplify.	$= 6t\sqrt{5}$

Example 5 Multiply and simplify: $\sqrt{x-7}\sqrt{x-7}$.

$$\begin{array}{rl}
\text{Multiply radicands.} & = \sqrt{(x-7)^2} \\
\text{Simplify.} & = x-7
\end{array}$$

Exercises Multiply and simplify.

1. $\sqrt{3}\sqrt{18}$

2. $\sqrt{5}\sqrt{60}$

3. $\sqrt{15}\sqrt{90}$

4. $\sqrt{18}\sqrt{14x}$

5. $\sqrt{12}\sqrt{18a}$

6. $\sqrt{13}\sqrt{13}$

7. $\sqrt{23}\sqrt{23y}$

8. $\sqrt{24w}\sqrt{40w}$

9. $\sqrt{2t}\sqrt{2t}$

10. $\sqrt{z+9}\sqrt{z+9}$

11. $\sqrt{7}\sqrt{4x+1}$

12. $\sqrt{125}\sqrt{200}$

13. $\sqrt{27x}\sqrt{6x}$

14. $\sqrt{x^5}\sqrt{x^8}$

15. $\sqrt{y^{12}}\sqrt{y^4}$

Integrated Review 17: COMPLETING THE SQUARE

A quadratic equation of the form $(x+c)^2 = d$ can be solved using the *principle of square roots*. Let's work through two examples.

$$x^2 = 81$$

$$x = \sqrt{81} \quad or \quad x = -\sqrt{81}$$

$$x = 9 \quad or \quad x = -9$$

The solutions are ± 9.

$$(x-8)^2 = 7$$

$$x - 8 = \sqrt{7} \quad or \quad x - 8 = -\sqrt{7}$$

$$x = 8 + \sqrt{7} \quad or \quad x = 8 - \sqrt{7}$$

The solutions are $8 \pm \sqrt{7}$.

Any quadratic equation, such as $x^2 - 10x + 3 = 0$, can be put in the form $(x+c)^2 = d$ by *completing the square*. We can add a number on both sides of an equation in order to make one side of the equation the square of a binomial. To determine the number to add, we take half the x-coefficient and square it.

Example 1
Solve: $x^2 - 10x + 3 = 0$ by completing the square.

Subtract 3 on both sides.	$x^2 - 10x = -3$
Add 25 on both sides.	$x^2 - 10x + 25 = -3 + 25$ Adding 25: $\frac{1}{2}(-10) = -5$, and $(-5)^2 = 25$
Express the left side as the square of a binomial and add on the right.	$(x-5)^2 = 22$
Use the principle of square roots.	$x - 5 = \sqrt{22} \quad or \quad x - 5 = -\sqrt{22}$
Simplify.	$x = 5 + \sqrt{22} \quad or \quad x = 5 - \sqrt{22}$
	The solutions are $5 \pm \sqrt{22}$.

Example 2
Solve: $x^2 + 3x - 2 = 0$ by completing the square.

Add 2 on both sides.	$x^2 + 3x = 2$
Add $\frac{9}{4}$ on both sides.	$x^2 + 3x + \frac{9}{4} = 2 + \frac{9}{4}$ Adding $\frac{9}{4}$: $\frac{1}{2} \cdot 3 = \frac{3}{2}$, and $\left(\frac{3}{2}\right)^2 = \frac{9}{4}$
Express the left side as the square of a binomial and add on the right.	$\left(x + \frac{3}{2}\right)^2 = \frac{8}{4} + \frac{9}{4}$ $\left(x + \frac{3}{2}\right)^2 = \frac{17}{4}$
Use the principle of square roots.	$x + \frac{3}{2} = \sqrt{\frac{17}{4}} \quad or \quad x + \frac{3}{2} = -\sqrt{\frac{17}{4}}$
Simplify.	$x = -\frac{3}{2} + \frac{\sqrt{17}}{2} \quad or \quad x = -\frac{3}{2} - \frac{\sqrt{17}}{2}$
	The solutions are $-\frac{3}{2} \pm \frac{\sqrt{17}}{2}$.

Check Your Understanding

Fill in the blanks with the number that completes the square on the left side.

1. $x^2 + 8x + \boxed{} = 3 + \boxed{}$

2. $y^2 - 6y + \boxed{} = -1 + \boxed{}$

3. $t^2 - 5t + \boxed{} = -4 + \boxed{}$

4. $x^2 + x + \boxed{} = 2 + \boxed{}$

Exercises Solve by completing the square.

1. $t^2 - 2t - 1 = 0$

2. $x^2 + 14x + 26 = 0$

3. $x^2 + 5x - 3 = 0$

4. $y^2 - y - 8 = 0$

5. $a^2 - 8a + 3 = 0$

6. $x^2 + 6x + 4 = 0$

7. $y^2 + 11y - 2 = 0$

8. $x^2 + \dfrac{3}{2}x - 5 = 0$

Integrated Review 18: FIND THE LCM OF ALGEBRAIC EXPRESSIONS

Equations containing rational expressions are called rational equations. To solve rational equations such as

$$\frac{5}{42} + \frac{7}{6} = \frac{x}{12}, \quad \text{and} \quad \frac{3}{x^2 - 4} - \frac{1}{5x + 10} = \frac{4}{x - 2},$$

the first step is to clear the equation of fractions. To do this, multiply all terms on both sides of the equation by the least common multiple, or LCM, of all the denominators. To find the LCM of two or more algebraic expressions, we factor them. Then we use each factor the greatest number of times that it occurs in any one factorization.

Example 1 Find the LCM of 42, 6 and 12.

$$\left.\begin{array}{l} 42 = 2 \cdot 3 \cdot 7 \\ 6 = 2 \cdot 3 \\ 12 = 2 \cdot 2 \cdot 3 \end{array}\right\} \text{The LCM is } 2 \cdot 2 \cdot 3 \cdot 7, \text{ or } 84.$$

Example 2 Find the LCM of $y^2 - 7y + 6$ and $y^2 - 5y - 6$.

$$\left.\begin{array}{l} y^2 - 7y + 6 = (y - 6)(y - 1) \\ y^2 - 5y - 6 = (y - 6)(y + 1) \end{array}\right\} \text{The LCM is } (y - 6)(y - 1)(y + 1).$$

Example 3 Find the LCM of $3x$, $6x^2$, and $15x + 18$.

$$\left.\begin{array}{l} 3x = 3 \cdot x \\ 6x^2 = 2 \cdot 3 \cdot x \cdot x \\ 15x + 18 = 3(5x + 6) \end{array}\right\} \text{The LCM is } 2 \cdot 3 \cdot x \cdot x \cdot (5x + 6), \text{ or } 6x^2(5x + 6).$$

Example 4 Find the LCM of $x^2 - 4$, $5x + 10$, and $x - 2$.

$$\left.\begin{array}{l} x^2 - 4 = (x + 2)(x - 2) \\ 5x + 10 = 5(x + 2) \\ x - 2 = x - 2 \end{array}\right\} \text{The LCM is } 5(x + 2)(x - 2).$$

Example 5 Find the LCM of $a^2 - 9$ and $a^2 - 6a + 9$.

$$\left.\begin{array}{l} a^2 - 9 = (a + 3)(a - 3) \\ a^2 - 6a + 9 = (a - 3)(a - 3) \end{array}\right\} \text{The LCM is } (a + 3)(a - 3)(a - 3), \text{ or } (a + 3)(a - 3)^2.$$

Example 6 Find the LCM of $4x^5 + 4x^4$ and $8x^4 - 8x^3 - 16x^2$.

$4x^5 + 4x^4 = 4x^4(x+1) = 2 \cdot 2 \cdot x \cdot x \cdot x \cdot x(x+1)$

$8x^4 - 8x^3 - 16x^2 = 8x^2(x^2 - x - 2) = 2 \cdot 2 \cdot 2 \cdot x \cdot x(x-2)(x+1)$

The LCM is $2 \cdot 2 \cdot 2 \cdot x \cdot x \cdot x \cdot x \cdot (x+1)(x-2)$, or $8x^4(x+1)(x-2)$.

Exercises Find the LCM by factoring.

1. $45, 54$

2. $12, 18, 48$

3. $3x^3, x, 9x^2$

4. $4x - 24, x + 6, x^2 - 36$

5. $a, a - 8, a + 8$

6. $y^2 - 7y - 30, y^2 + 13y + 30$

7. $x^2 - 4, x^2 + 5x + 6$

8. $5t, t^2, 10t - 40, 10t^3$

9. $z^7, z^2 - 1, z^3 - 2z^2 + z$

10. $a^2 - 4a - 21, a^2 - 6a - 7$

11. $30x^2 - 30, 105x - 105$

12. $9y^2, 9y^3, 24y - 27$

13. $9q^7 + 9q^6, 6q^6 - 30q^5 - 36q^4$

14. $35c^2, c + 7, 70c, c^2 + 2c - 35$

Integrated Review 19:   RAISING RADICALS TO POWERS

A **radical equation** has a variable in one or more radicands. For example,
$$\sqrt{x} = 2, \quad \sqrt[3]{4x} - 1 = 7, \text{ and } \sqrt{x+2} = \sqrt{x+3} + 8$$
are radical equations. To solve a radical equation, we raise the radical(s) to a power.

Example 1 Simplify: $\left(\sqrt{3}\right)^2$.

Write $\left(\sqrt{3}\right)^2$ as a product.	$\sqrt{3} \cdot \sqrt{3}$
Multiply.	$= \sqrt{3^2}$, or $\sqrt{9}$
Simplify.	$= 3$

Example 2 Simplify: $\left(\sqrt[3]{x}\right)^3$.

Write $\left(\sqrt[3]{x}\right)^3$ as a product.	$\sqrt[3]{x} \cdot \sqrt[3]{x} \cdot \sqrt[3]{x}$
Multiply.	$= \sqrt[3]{x^3}$
Simplify.	$= x$

Example 3 Simplify: $\left(\sqrt{2y-5}\right)^2$.

Write $\left(\sqrt{2y-5}\right)^2$ as a product.	$\sqrt{2y-5} \cdot \sqrt{2y-5}$
Multiply.	$= \sqrt{(2y-5)^2}$
Simplify.	$= 2y - 5$

In general, we see that $\left(\sqrt[n]{a}\right)^n = \sqrt[n]{a^n} = a$, where n is an integer and $n \geq 2$. (Here we assume that no radicands are formed by raising negative quantities to even powers.)

When we have an equation that can be written in the form $\sqrt{x} + a = \sqrt{y}$, we use FOIL to square the expression $\sqrt{x} + a$.

Example 4 Simplify: $\left(\sqrt{x} - 3\right)^2$.

Use FOIL.	$= \left(\sqrt{x}\right)^2 - 2 \cdot \sqrt{x} \cdot 3 + 3^2$
Simplify.	$= x - 6\sqrt{x} + 9$

61

Example 5 Simplify: $\left(\sqrt{y-2}+5\right)^2$.

Use FOIL.	$=\left(\sqrt{y-2}\right)^2+2\cdot\sqrt{y-2}\cdot 5+5^2$
Simplify.	$=y-2+10\sqrt{y-2}+25$
Collect like terms.	$=y+23+10\sqrt{y-2}$

Example 6 Simplify: $\left(4-\sqrt{2z+1}\right)^2$.

Use FOIL.	$=4^2-2\cdot 4\sqrt{2z+1}+\left(\sqrt{2z+1}\right)^2$
Simplify.	$=16-8\sqrt{2z+1}+2z+1$
Collect like terms.	$=17-8\sqrt{2z+1}+2z$

Exercises Simplify.

1. $\left(\sqrt{6}\right)^2$

2. $\left(\sqrt[3]{2}\right)^3$

3. $\left(\sqrt{x+7}\right)^2$

4. $\left(\sqrt{y-8}\right)^2$

5. $\left(\sqrt[3]{3z+1}\right)^3$

6. $\left(\sqrt{x}+2\right)^2$

7. $\left(4-\sqrt{3x}\right)^2$

8. $\left(\sqrt{y-4}+1\right)^2$

9. $\left(3-\sqrt{2t+5}\right)^2$

Integrated Review 20: INTRODUCTION TO POLYNOMIALS

A **monomial** is a constant or a constant times some variable or variables raised to powers that are nonnegative integers. A **polynomial** is a monomial or a combination of sums and/or differences of monomials.

The following are examples of *monomials*:

$$7, \quad z, \quad \frac{1}{2}x, \quad 4a^2, \quad -2.8y^3, \quad 5ab^4.$$

Expressions like these are *polynomials*:

$$-6, \quad t-9, \quad x^2+5x+6, \quad 5x^3+\frac{3}{4}x^2-x, \quad c^2-d^2.$$

The following are algebraic expressions that are *not* polynomials:

$$(1) \ \ \frac{y-2}{y+3}, \qquad (2) \ \ x^5-x^3+\frac{1}{x}, \qquad (3) \ \ w^{1/2}.$$

Expression (1) is not a polynomial because it represents a quotient. In expression (2), although $1/x$ can be written as x^{-1}, this is not a monomial because the exponent is negative. The exponent in expression (3) is not a nonnegative integer, so $w^{1/2}$ is not a monomial.

In this lesson, we will consider only polynomials in one variable.

Exercises Determine if the expression is a polynomial. Answer yes or no.

1. $4y-\dfrac{1}{y}$

2. 9

3. $2x^2+4x-5$

4. $8a^3$

5. $z^{1/2}+10$

6. $\dfrac{t}{t+6}$

The **terms** of a polynomial are separated by $+$ signs. The polynomial

$$5x^4-2x^3-x+8, \text{ or } 5x^4+\left(-2x^3\right)+\left(-x\right)+8,$$

has four terms:

$$5x^4, \qquad -2x^3, \qquad -x, \qquad \text{and} \quad 8.$$

The **coefficients** of the terms are 5, -2, -1, and 8. The term 8 is called a **constant term**.

The **degree of a term** is the exponent of the variable or the sum of the exponents of the variables, if there are variables For example,

the degree of the term $4y^7$ is 7,

the degree of the term $3x^2y^3$ is $2+3$, or 5, and

the degree of the term 9, or $9x^0$, is 0.

Because we can express 0 as $0 = 0x^5 = 0x^8$, and so on, using any exponent we wish, the term 0 has *no* degree.

The **degree of a polynomial** is the same as the degree of its term of highest degree. For example,

the degree of the polynomial $2 - 9x^2 + x^6 + 4x$ is 6.

The **leading term** of a polynomial is the term of highest degree. Its coefficient is called the **leading coefficient**. For example,

the leading term of $9x^2 - 5x^3 + x - 10$ is $-5x^3$ and

the leading coefficient is -5.

Example 1 Identify the terms, the degree of each term, and the degree of the polynomial $2x^3 + 8x^2 - 17x - 3$. Then identify the leading term, the leading coefficient, and the constant term.

Term	$2x^3$	$8x^2$	$-17x$	-3
Degree of Term	3	2	1	0
Degree of Polynomial	3			
Leading Term	$2x^3$			
Leading Coefficient	2			
Constant Term	-3			

Exercises Identify the terms, the degree of each term, and the degree of the polynomial. Then identify the leading term, the leading coefficient, and the constant term.

7. $3y^4 - 6y^3 + 8y^2 - \dfrac{2}{3}y + 5$

Term				
Degree of Term				
Degree of Polynomial				
Leading Term				
Leading Coefficient				
Constant Term				

8. $x - 3x^4 + 4x^3 - 13 + 7x^2 - 6x^5$

Term					
Degree of Term					
Degree of Polynomial					
Leading Term					
Leading Coefficient					
Constant Term					

The following are some names for certain types of polynomials.

Type	Definition	Examples
Monomial	One term	$17, \ -4x^5$
Binomial	Two terms	$a^2 + 10, \ w - 1$
Trinomial	Three terms	$y^2 - 6y + 8, \ b^6 + 4b^3 - 9b$

Exercises Classify the polynomial as a monomial, a binomial, or a trinomial.

9. $x^2 - 36$

10. $w^2 - 7w - 44$

11. -18

12. $5 - t^2 + t^8$

13. y

14. $q^4 + 16$

We generally arrange polynomials in one variable in **descending order** so that the exponents *decrease* from left to right. Sometimes they may be written so that the exponents *increase* from left to right, which is **ascending order**. In general, if an exercise is written in a particular order, we write the answer in that same order.

Example 2 Consider $12 + x^2 - 7x$. Arrange in descending order and then in ascending order.

Descending order: $x^2 - 7x + 12$ Ascending order: $12 - 7x + x^2$

Example 3 Consider $y^4 + 5y - 20 - 2y^5 - 3y^2$. Arrange in descending order and then in ascending order.

Descending order: $-2y^5 + y^4 - 3y^2 + 5y - 20$

Ascending order: $-20 + 5y - 3y^2 + y^4 - 2y^5$

Exercises Arrange the polynomial in descending order and then in ascending order.

15. $y + 2y^3 - 8y^2$

16. $3 - a^4 + a^3$

17. $-5t + 3t^2 - t^3 + 3$

18. $12x + 5 + 8x^5 - 4x^3$

19. $23 - 9b^3 + 4b - 5b^2$

20. $\dfrac{3}{5}q + 5q^7 - \dfrac{1}{3}$

Notes:

Integrated Review 21: FACTORING BY GROUPING

In some polynomials, a *binomial* is a common factor.

Example 1 Factor: $y^2(y+2)+3(y+2)$.

The common factor is the binomial $y+2$. We factor out the common factor, $y+2$.

$$y^2(y+2)+3(y+2)=(y+2)(y^2+3)$$

In Example 2, we factor a polynomial with four terms. We begin by grouping pairs of terms.

Example 2   Factor: $x^3+5x^2+7x+35$.

Group before factoring.	$=(x^3+5x^2)+(7x+35)$
Factor each binomial.	$=x^2(x+5)+7(x+5)$
Factor out the common factor, $x+5$.	$=(x+5)(x^2+7)$

This is called **factoring by grouping**.

Example 3   Factor: $5t^3-15t^2-t+3$.

Consider the first two terms.	$5t^3-15t^2$
Factor out the greatest common factor, or GCF, $5t^2$.	$=5t^2(t-3)$
Consider the last two terms.	$-t+3$
Factor out -1 so one of the factors is $t-3$. Note: $-t+3=-1(t-3)$.	$=-1(t-3)$

Now we can factor out $t-3$ in the polynomial, as shown below.

	$5t^3-15t^2-t+3$
Group before factoring.	$=(5t^3-15t^2)+(-t+3)$
Factor each binomial. Note: $-t+3=-1(t-3)$.	$=5t^2(t-3)-1(t-3)$
Factor out the common factor, $t-3$.	$=(t-3)(5t^2-1)$

Example 4   Factor: $2a^3 + 12a^2 - 5a - 30$.

Group before factoring.	$= (2a^3 + 12a^2) + (-5a - 30)$
Factor each binomial. Note: $-5a - 30 = -5(a + 6)$.	$= 2a^2(a + 6) - 5(a + 6)$
Factor out the common factor, $a + 6$.	$= (a + 6)(2a^2 - 5)$

Example 5   Factor: $4z^3 - 9 + z - 36z^2$.

Arrange in descending order.	$= 4z^3 - 36z^2 + z - 9$
Group before factoring.	$= (4z^3 - 36z^2) + (z - 9)$
Factor each binomial.	$= 4z^2(z - 9) + 1(z - 9)$
Factor out the common factor, $z - 9$.	$= (z - 9)(4z^2 + 1)$

Example 6   Factor: $3x^7 - 15x^6 + 6x - 30$.

Factor out the common factor, 3.	$= 3(x^7 - 5x^6 + 2x - 10)$
Group before factoring.	$= 3[(x^7 - 5x^6) + (2x - 10)]$
Factor each binomial.	$= 3[x^6(x - 5) + 2(x - 5)]$
Factor out the common factor, $x - 5$.	$= 3(x - 5)(x^6 + 2)$

Check Your Understanding

Not all polynomials with four terms can be factored by grouping. For example, $x^3 + x^2 + 3x - 3 = x^2(x + 1) + 3(x - 1)$ and cannot be factored by grouping.

Determine if the polynomial can be factored by grouping. Answer with "yes" or "no".

1. $y^3 + 8y^2 - 5y - 40$ 2. $a^3 - 2a^2 + 5a + 10$

3. $2x^3 + 3x^2 - 8x + 12$ 4. $t^3 - 6t + 9t^2 - 54$

Exercises Factor by grouping, if possible. In Exercises 1 and 2, fill-in the blanks in key steps.

1. $y^3 + 4y^2 + 9y + 36$

   $= \left(y^3 + 4y^2\right) + \left(9y + 36\right)$

   $= \boxed{}(y+4) + 9\left(\boxed{}\right)$

   $= \left(\boxed{}\right)\left(y^2 + 9\right)$

2. $3s^3 - 15s^2 - 7s + 35$

   $= \left(3s^3 - 15s^2\right) + \left(\boxed{}\right)$

   $= 3s^2\left(\boxed{}\right) + \boxed{}(s - 5)$

   $= (s - 5)\left(\boxed{}\right)$

3. $c^3 + 6c^2 - 5c - 30$

4. $x^3 - 7x^2 + 4x - 28$

5. $3x^3 - 30x^2 - x - 10$

6. $18y - 2y^2 - 18 + 2y^3$

7. $x^3 - 5x^2 + x - 5$

8. $2y^3 + 3y^2 - 6y - 9$

9. $xy - xz + wy - wz$

10. $2x^4 + 6x^2 - 5x^2 - 15$

Notes:

Integrated Review 22: INTEGERS AS EXPONENTS

Definitions and Rules for Exponents

Exponent of 1:	$a^1 = a$
Exponent of 0:	$a^0 = 1, \ a \neq 0$
Negative exponents:	$a^{-n} = \dfrac{1}{a^n}, \ \dfrac{1}{a^{-n}} = a^n, \ a \neq 0$
Product Rule:	$a^m \cdot a^n = a^{m+n}$
Quotient Rule:	$\dfrac{a^m}{a^n} = a^{m-n}, \ a \neq 0$
Power Rule:	$\left(a^m\right)^n = a^{mn}$
Raising a product to a power:	$\left(ab\right)^n = a^n b^n$
Raising a quotient to a power:	$\left(\dfrac{a}{b}\right)^n = \dfrac{a^n}{b^n}, \ b \neq 0$

Examples Simplify.

1. $5^0 = 1$

2. $18^1 = 18$

3. $y^1 = y$

4. $t^0 = 1, \ t \neq 0$

Examples Express using positive exponents. Then simplify.

5. $a^{-6} = \dfrac{1}{a^6}$

6. $\dfrac{1}{4^{-2}} = 4^2 = 16$

7. $\dfrac{1}{x^{-3}} = x^3$

8. $3^{-4} = \dfrac{1}{3^4} = \dfrac{1}{81}$

Examples Simplify using the product rule.

9. $8^5 \cdot 8^2 = 8^{5+2} = 8^7$

10. $b^{-4} \cdot b^7 = b^{-4+7} = b^3$

11. $x \cdot x^{12} = x^{1+12} = x^{13}$

12. $2^{-10} \cdot 2^3 = 2^{-10+3} = 2^{-7} = \dfrac{1}{2^7}$

Examples Simplify using the quotient rule.

13. $\dfrac{a^8}{a^3} = a^{8-3} = a^5$

14. $\dfrac{7^2}{7^5} = 7^{2-5} = 7^{-3} = \dfrac{1}{7^3}$

15. $\dfrac{y^6}{y^{10}} = y^{6-10} = y^{-4} = \dfrac{1}{y^4}$

16. $\dfrac{13^9}{13^4} = 13^{9-4} = 13^5$

Examples Simplify using the power rule.

17. $\left(9^2\right)^6 = 9^{2 \cdot 6} = 9^{12}$

18. $\left(x^{-3}\right)^5 = x^{-3 \cdot 5} = x^{-15} = \dfrac{1}{x^{15}}$

19. $\left(3^2\right)^{-3} = 3^{2(-3)} = 3^{-6} = \dfrac{1}{3^6}$

20. $\left(y^4\right)^8 = y^{4 \cdot 8} = y^{32}$

Examples Simplify using the rule for raising a product to a power.

21. $\left(3^5 y^2\right)^4 = \left(3^5\right)^4 \cdot \left(y^2\right)^4 = 3^{5 \cdot 4} \cdot y^{2 \cdot 4} = 3^{20} y^8$

22. $\left(2z^6\right)^3 = 2^3 \cdot \left(z^6\right)^3 = 8 \cdot z^{6 \cdot 3} = 8z^{18}$

Examples Simplify using the rule for raising a quotient to a power.

23. $\left(\dfrac{2}{5}\right)^3 = \dfrac{2^3}{5^3} = \dfrac{8}{125}$

24. $\left(\dfrac{x^2}{4^5}\right)^6 = \dfrac{\left(x^2\right)^6}{\left(4^5\right)^6} = \dfrac{x^{2 \cdot 6}}{4^{5 \cdot 6}} = \dfrac{x^{12}}{4^{30}}$

Exercises Simplify. Express answer using positive exponents.

1. 3^{-5}

2. $a^7 a^{11}$

3. $\dfrac{x^{16}}{x^4}$

4. $\left(2^7\right)^3$

5. $\left(\dfrac{5}{3}\right)^3$

6. 100^1

7. 3^0

8. $x^3 \cdot x^{-5}$

9. $\dfrac{1}{y^{-10}}$

10. $\left(7^2 x^4\right)^3$

11. $4^2 \cdot 4^9$

12. $\dfrac{1}{3^{-4}}$

13. $\left(t^3\right)^{-4}$

14. z^{-2}

15. $\dfrac{2^4}{2^9}$

16. $\left(s^2\right)^3$

17. p^1

18. $y^0, \; y \neq 0$

19. $\left(\dfrac{y^3}{2^5}\right)^4$

20. $\dfrac{10^{15}}{10^5}$

21. $w \cdot w^8$

22. $\dfrac{c^{-4}}{c^{-11}}$

23. $\left(3x^2\right)^2$

24. $y^{-2} \cdot y^{-4}$

Integrated Review 23: INTRODUCTION TO LOGARITHMS

Logarithmic functions are inverses of exponential functions and have applications in fields such as business, science, psychology, and sociology.

Logarithmic Function, Base a

- We define $y = \log_a x$ as that number y such that $x = a^y$, where $x > 0$ and a is a positive constant other than 1.

$$\log_a x = y \text{ is equivalent to } x = a^y.$$

- A logarithm is an EXPONENT!
- We read $\log_a x$ as "the logarithm, base a, of x."

Example 1 Find $\log_2 8$.

We read $\log_2 8$ as "the logarithm, base 2, of 8."

Let $y = \log_2 8$. This equation is equivalent to $2^y = 8$. The power to which we raise 2 to get 8 is 3; thus $\log_2 8 = 3$. Check: $2^3 = 8$.

Example 2 Find $\log_{10} 100,000$.

Let $y = \log_{10} 100,000$, or $10^y = 100,000$. The power to which we raise 10 to get 100,000 is 5; thus $\log_{10} 100,000 = 5$. Check: $10^5 = 100,000$.

Base-10 logarithms are called **common logarithms**. The abbreviation **log**, with no *base* written, is used for the common logarithm. Thus, $\log a = \log_{10} a$. For example, $\log 100,000 = \log_{10} 100,000$.

It is helpful to review some definitions and rules of exponents.

- For any real number a that is nonzero and any integer n,

$$a^{-n} = \frac{1}{a^n}. \qquad \left(\text{Example: } 9^{-2} = \frac{1}{9^2} \right)$$

- $a^1 = a$, for any number a. $\qquad \left(\text{Example: } 4^1 = 4 \right)$
- $a^0 = 1$, for any nonzero number a. $\qquad \left(\text{Example: } 8^0 = 1 \right)$

Example 3 Find $\log 0.0001$. $\left(\text{Remember: } \log 0.0001 = \log_{10} 0.0001 \right)$

$$0.0001 = \frac{1}{10,000} = \frac{1}{10^4} = 10^{-4}$$

$$\log 0.0001 = \log 10^{-4}$$

We let $y = \log_{10} 10^{-4}$, or $10^y = 10^{-4}$. The power to which we raise 10 to get 0.0001 is -4; thus $\log 0.0001 = -4$.

Example 4 Find $\log_5 \dfrac{1}{25}$.

$$\dfrac{1}{25} = \dfrac{1}{5^2} = 5^{-2}$$

$$\log_5 \dfrac{1}{25} = \log_5 5^{-2}$$

We let $y = \log_5 5^{-2}$, or $5^y = 5^{-2}$. The power to which we raise 5 to get $\dfrac{1}{25}$ is -2; thus

$\log_5 \dfrac{1}{25} = -2$.

Example 5 Find $\log_3 1$.

Let $y = \log_3 1$, or $3^y = 1$. We know that $3^0 = 1$. Substituting 3^0 for 1 in $3^y = 1$, we have $3^y = 3^0$. The power to which we raise 3 to get 1 is 0; thus $\log_3 1 = 0$.

Example 6 Find $\log_7 7$.

Let $y = \log_7 7$, or $7^y = 7$. We know that $7^1 = 7$. Substituting 7^1 for 7 in $7^y = 7$, we have $7^y = 7^1$. The power to which we raise 7 to get 7 is 1; thus $\log_7 7 = 1$.

- $\log_a 1 = 0$ The logarithm, base a, of 1 is always 0.
- $\log_a a = 1$ The logarithm, base a, of a is always 1.

Exercises Find each of the following. Remember: a logarithm is an exponent.

1. $\log_6 36$

2. $\log 1000$

3. $\log_{19} 19$

4. $\log_8 1$

5. $\log \dfrac{1}{10}$

6. $\log 1,000,000$

7. $\log_4 4$

8. $\log_5 625$

9. $\log 0.01$

10. $\log_7 \dfrac{1}{49}$

11. $\log 0.00001$

12. $\log 1$

Interactive Preview Worksheets

Correlation Guide

The **Preview Worksheets** in this Notebook accompany *College Algebra*, 5th edition, by Beecher/Penna/Bittinger. The following table contains a correlation between the worksheets and the sections in the text.

Preview Worksheet #	Section in the Text
1	1.2
2	2.1
3	2.5
4	3.3
5	3.4
6	3.4
7	3.5
8	4.1
9	4.5
10	5.1
11	5.5
12	5.5
13	6.5
14	7.3
15	7.2
16	7.3

Interactive Preview 1: FUNCTION VALUES; DOMAIN AND RANGE

Example 1 Consider the graph of $f(x) = x^2 + 3x - 4$. Find $f(2)$ and $f(-3)$.

To find the function value $f(2)$ from the graph, we locate the input 2 on the x-axis, move vertically to the graph of the function, and then move horizontally to find the output on the y-axis. We see that $f(2) = 6$. The ordered pair label for the point on the graph is $(2, 6)$.

To find the function value $f(-3)$ from the graph, we locate the input -3 on the x-axis, move vertically to the graph of the function, and then move horizontally to find the output on the y-axis. We see that $f(-3) = -4$. The ordered-pair label for the point on the graph is $(-3, -4)$. We also see that $f(0) = -4$, $f(1) = 0$, and $f(-5) = 6$.

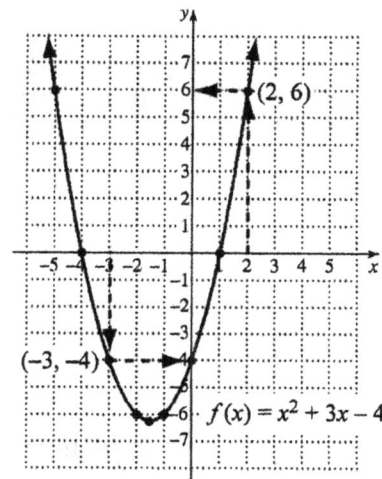

$f(x) = x^2 + 3x - 4$

Exercises

Use the graph of the function to determine function values and fill in the coordinates of the ordered-pair labels.

1.

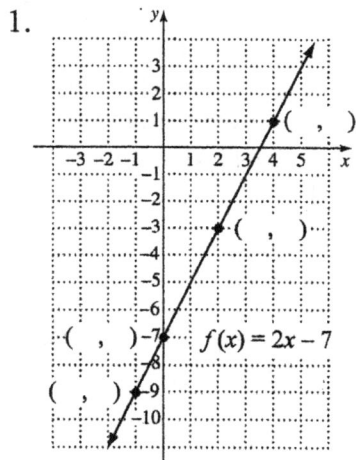

$f(x) = 2x - 7$

$f(2) =$ _____

$f(-1) =$ _____

$f(0) =$ _____

$f(4) =$ _____

2.

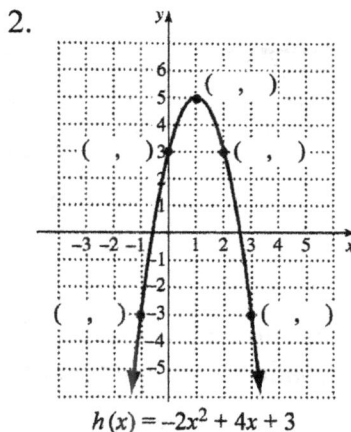

$h(x) = -2x^2 + 4x + 3$

$h(0) =$ _____

$h(1) =$ _____

$h(-1) =$ _____

$h(3) =$ _____

$h(2) =$ _____

3.

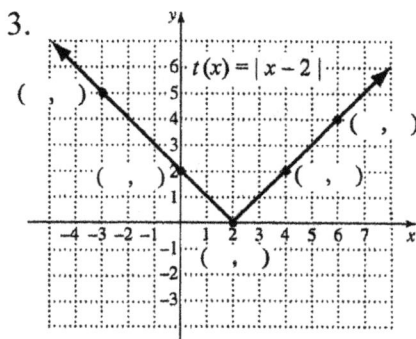

$t(x) = |x - 2|$

$t(-3) =$ _____

$t(4) =$ _____

$t(0) =$ _____

$t(2) =$ _____

$t(6) =$ _____

4.

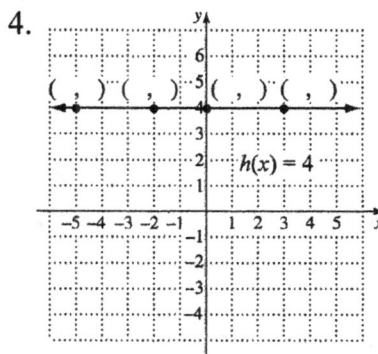

$h(x) = 4$

$h(3) =$ _____

$h(-2) =$ _____

$h(-5) =$ _____

$h(0) =$ _____

5.

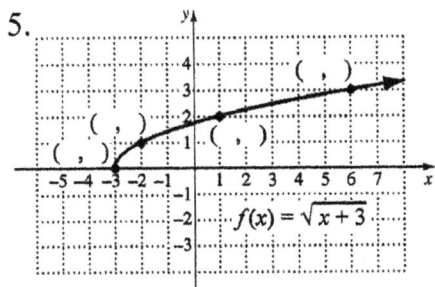

$f(-3) = $ _____

$f(-2) = $ _____

$f(1) = $ _____

$f(6) = $ _____

$f(x) = \sqrt{x+3}$

6.

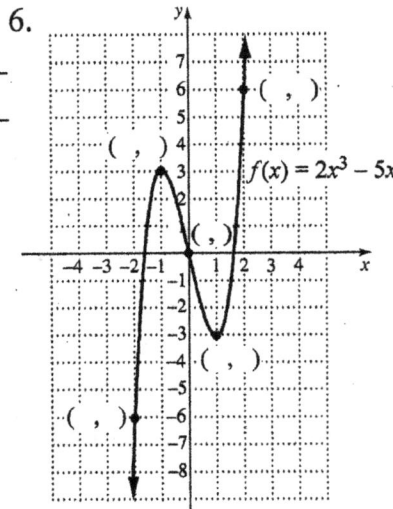

$f(x) = 2x^3 - 5x$

$f(-1) = $ _____

$f(2) = $ _____

$f(-2) = $ _____

$f(0) = $ _____

$f(1) = $ _____

Example 2 Using the graph, find the domain and the range of the function. Shade the domain (inputs) on the x-axis and the range (outputs) on the y-axis.

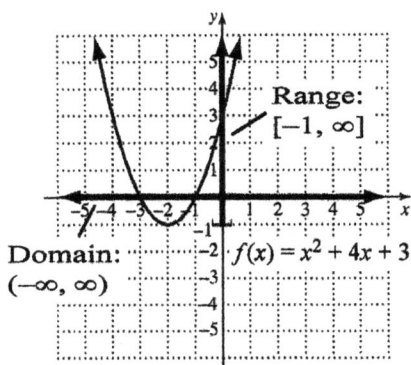

Range: [−1, ∞]

Domain: (−∞, ∞)

$f(x) = x^2 + 4x + 3$

To determine the domain of a function, we look for inputs on the x-axis that correspond to a point on the graph. We see that they include the entire set of real numbers. The domain is $(-\infty, \infty)$. To find the range, we look for outputs on the y-axis that correspond to a point on the graph. We see that they include −1 and all real numbers greater than −1. The bracket at −1 indicates that −1 is included in the interval. The range is $\{y \mid y \geq -1\}$, or $[-1, \infty)$.

Exercises

Using the graphs of the functions shown in Exercises 1-6, find the domain and the range of each function (Hint: Shade the domain and the range of the function on the x- and y- axes).

7. $f(x) = 2x - 7$
   (See Exercise 1.)

   Domain: _____
   Range: _____

8. $h(x) = -2x^2 + 4x + 3$
   (See Exercise 2.)

   Domain: _____
   Range: _____

9. $t(x) = |x - 2|$
   (See Exercise 3.)

   Domain: _____
   Range: _____

10. $h(x) = 4$
    (See Exercise 4.)

    Domain: _____
    Range: _____

11. $f(x) = \sqrt{x+3}$
    (See Exercise 5.)

    Domain: _____
    Range: _____

12. $f(x) = 2x^3 - 5x$
    (See Exercise 6.)

    Domain: _____
    Range: _____

Interactive Preview 2: GRAPHING PIECEWISE FUNCTIONS

When a function is defined by different equations for various parts of its domain, it is said to be defined **piecewise**.

Example 1 Graph the function

$$f(x) = \begin{cases} \dfrac{1}{2}x, \text{ if } x \le -1 & \text{Equation 1} \\[2mm] x+3, \text{ if } x > -1 & \text{Equation 2} \end{cases}$$

This function is defined using two different equations: $f(x) = \dfrac{1}{2}x$ and $f(x) = x+3$. We graph each equation separately. Let's visualize the domain of each.

Domain

Equation 1: $(-\infty, -1]$

Equation 2: $(-1, \infty)$

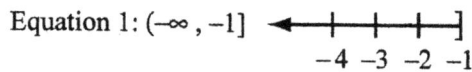

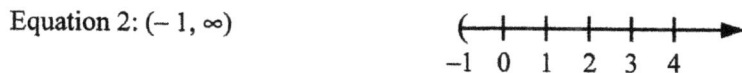

Equation 1: Choose x-values less than or equal to -1.

If $x = -1$, $y = \dfrac{1}{2}(-1) = -\dfrac{1}{2}$.

If $x = -2$, $y = \dfrac{1}{2}(-2) = -1$.

If $x = -4$, $y = \dfrac{1}{2}(-4) = -2$.

$x \le -1$	$f(x) = \dfrac{1}{2}x$
-1	$-\dfrac{1}{2}$
-2	-1
-4	-2

Equation 2: Choose x-values greater than -1.

If $x = 0$, $y = 0+3 = 3$.
If $x = 1$, $y = 1+3 = 4$.
If $x = 2$, $y = 2+3 = 5$.

$x > -1$	$f(x) = x+3$
0	3
1	4
2	5

-1 is in the domain of $f(x) = \frac{1}{2}x$. We use a solid dot at $(-1, -\frac{1}{2})$.

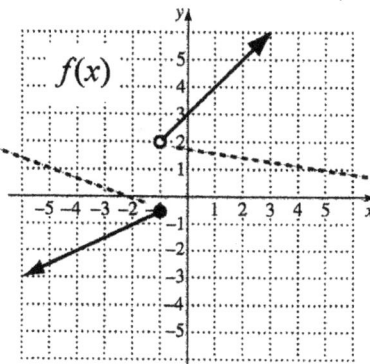

For $f(x) = x + 3$, if $x = -1$, $f(-1) = -1 + 3 = 2$.
But -1 is **not** in the domain of $f(x) = x + 3$.
Thus we use an open circle at $(-1, 2)$.

Example 2 Graph the function

$$g(x) = \begin{cases} x^2, & \text{if } x < 0 & \text{Equation 1} \\ -2, & \text{if } 0 \le x < 3 & \text{Equation 2} \\ 4-x, & \text{if } x \ge 3 & \text{Equation 3} \end{cases}$$

Domain

Equation 1: $(-\infty, 0)$

Equation 2: $[0, 3)$

Equation 2: $[3, \infty)$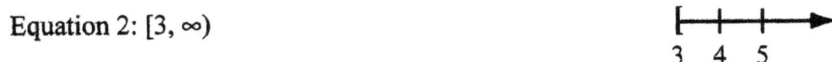

Equation 1

$x < 0$	$g(x) = x^2$
$-\dfrac{1}{2}$	$\dfrac{1}{4}$
-1	1
-2	4

- For $g(x) = x^2$, if $x = 0$, $g(0) = 0^2 = 0$. But 0 **is not** in the domain of $g(x) = x^2$. Thus, we use an open circle at $(0, 0)$.

Equation 2

$0 \le x < 3$	$g(x) = -2$
0	-2
1	-2
2	-2

- 0 **is** in the domain of $g(x) = -2$. We use a solid dot at $(0, -2)$.
- For $g(x) = -2$, if $x = 3$, $g(3) = -2$. But 3 **is not** in the domain of $g(x) = -2$. Thus, we use an open circle at $(3, -2)$.

Equation 3

$x \ge 3$	$g(x) = 4-x$
3	1
4	0
5	-1

- 3 **is** in the domain of $g(x) = 4-x$. We use a solid dot at $(3, 1)$.

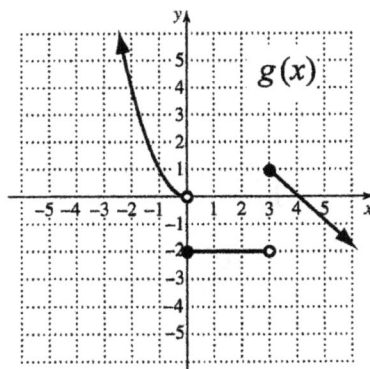

Exercises Graph the piecewise function. In Exercises 1 and 2, fill in the blanks at key steps in the graphing process. In Exercises 3-6, you need to show only the graph.

1. $f(x) = \begin{cases} 3x, & \text{if } x < 2 \quad (1) \\ x-1, & \text{if } x \geq 2 \quad (2) \end{cases}$

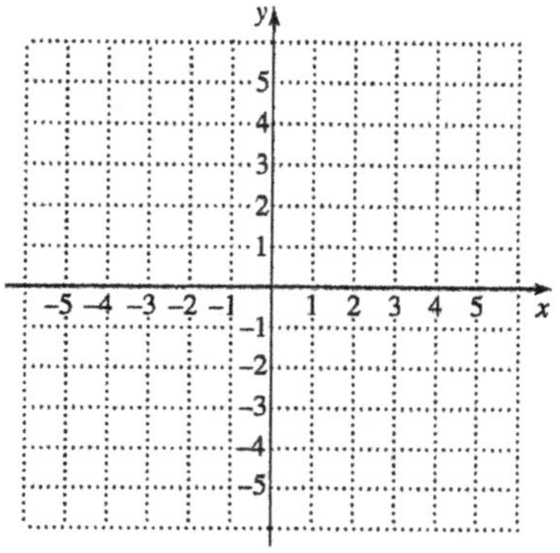

Domain of Equation 1: $\left(\boxed{}, 2 \right)$

Domain of Equation 2: $\left[\boxed{}, \infty \right)$

Equation 1

$x < 2$	$f(x) = 3x$
1	$\boxed{}$
0	$\boxed{}$
−1	$\boxed{}$

Equation 2

$x \geq 2$	$f(x) = x-1$
2	$\boxed{}$
4	$\boxed{}$
5	$\boxed{}$

- If $f(x) = 3x$ and $x = 2$, $f(2) = 3 \cdot \boxed{} = 6$. Since 2 _____ in the domain of
  is / is not
  $f(x) = 3x$, we use a(an) _____ at $\left(2, \boxed{}\right)$.
  solid dot / open circle

- If $f(x) = x-1$ and $x = 2$, $f(2) = \boxed{} - 1 = 1$. Since 2 _____ in the domain
  is / is not
  of $f(x) = x-1$ we use a(an) _____ at $\left(2, \boxed{}\right)$.
  solid dot / open circle

2. $g(x) = \begin{cases} -x+1, & \text{if } x < -1 \quad (1) \\ -3, & \text{if } -1 \leq x < 4 \quad (2) \\ 2x-4, & \text{if } x \geq 4 \quad (3) \end{cases}$

Domain of Equation 1: $\left(-\infty, \boxed{} \right)$

Domain of Equation 2: $\left[\boxed{}, \boxed{} \right)$

Domain of Equation 3: $\left[4, \boxed{} \right)$

Equation 1

$x < -1$	$g(x) = -x+1$
−2	$\boxed{}$
−3	$\boxed{}$
−4	$\boxed{}$

Equation 2

$-1 \leq x < 4$	$g(x) = -3$
−1	$\boxed{}$
1	$\boxed{}$
3	$\boxed{}$

Equation 3

$x \geq 4$	$g(x) = 2x-4$
4	$\boxed{}$
5	$\boxed{}$
6	$\boxed{}$

- If $g(x) = -x + 1$ and $x = -1$, $g(-1) = -\left(\boxed{}\right) + 1 = 2$.

  Since -1 _____ in the domain of $g(x) = -x + 1$,

        is / is not

  we use a(an) _____ at $\left(-1, \boxed{}\right)$.

       solid dot / open circle

- If $g(x) = -3$ and $x = -1$, $g(-1) = -3$. Since -1

  _____ in the domain of $g(x) = -3$ we use a(an)

  is / is not

  _____ at $\left(-1, \boxed{}\right)$.

  solid dot / open circle

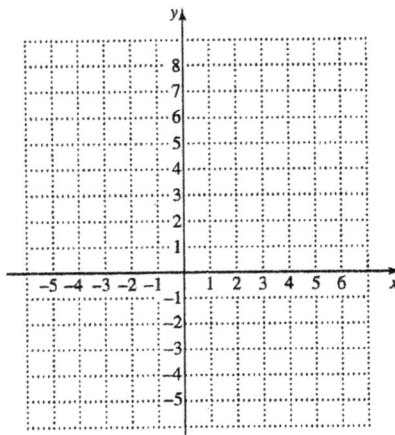

- If $g(x) = -3$ and $x = 4$, $g(4) = -3$. Since 4

  _____ in the domain of $g(x) = -3$ we use a(an)

  is / is not

  _____ at $\left(4, \boxed{}\right)$.

  solid dot / open circle

- If $g(x) = 2x - 4$ and $x = 4$, $g(4) = 2 \cdot \boxed{} - 4 = 4$.

  Since 4 _____ in the domain of $g(x) = 2x - 4$ we

        is / is not

  use a(an) _____ at $\left(4, \boxed{}\right)$.

       solid dot / open circle

3. $h(x) = \begin{cases} 4, & \text{if } x \le 0 \quad (1) \\ 2 - x, & \text{if } x > 0 \quad (2) \end{cases}$

Domain of Equation 1: _____

Domain of Equation 2: _____

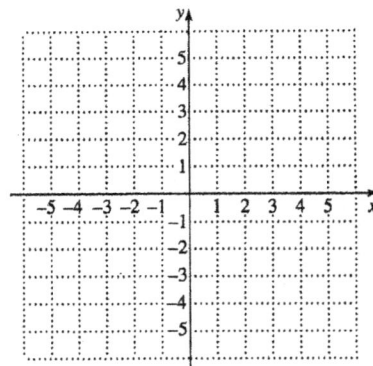

Equation 1		Equation 2	
$x \le 0$	$h(x) = 4$	$x > 0$	$h(x) = 2 - x$

4. $f(x) = \begin{cases} x^2, & \text{if } x \leq -1 \quad (1) \\ 2x - 3, & \text{if } x > -1 \quad (2) \end{cases}$

Domain of Equation 1: _____

Domain of Equation 2: _____

Equation 1			Equation 2	
$x \leq -1$	$f(x) = x^2$		$x > -1$	$f(x) = 2x - 3$

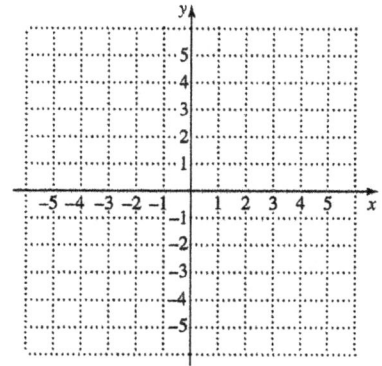

5. $t(x) = \begin{cases} |x|, & \text{if } x < -2 \quad (1) \\ \dfrac{1}{2}x - 1, & \text{if } x \geq -2 \quad (2) \end{cases}$

Domain of Equation 1: _____

Domain of Equation 2: _____

Equation 1			Equation 2	
x	$t(x)$		x	$t(x)$

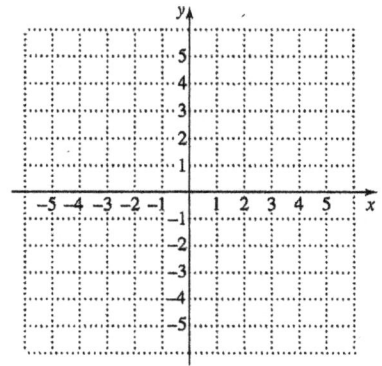

6. $g(x) = \begin{cases} -x, & \text{if } x < -5 \quad \text{(1)} \\ x+3, & \text{if } -5 \le x < 2 \quad \text{(2)} \\ \dfrac{1}{4}x^2 - 6, & \text{if } x \ge 2 \quad \text{(3)} \end{cases}$

Domain of Equation 1: _____

Domain of Equation 2: _____

Domain of Equation 3: _____

Equation 1		Equation 2		Equation 3	
x	$g(x)$	x	$g(x)$	x	$g(x)$

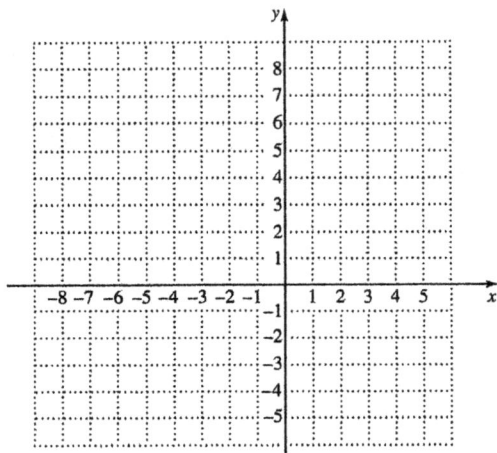

Notes:

Interactive Preview 3: TRANSFORMATIONS

The squaring function $f(x) = x^2$ is shown below. We can create graphs of additional functions by

- shifting the graph of $f(x) = x^2$ horizontally or vertically,
- reflecting the graph of $f(x) = x^2$ across an axis, and
- stretching or shrinking the graph of $f(x) = x^2$.

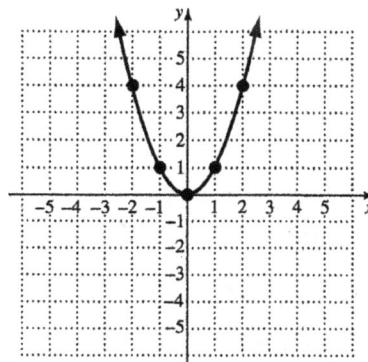

The new graphs are **transformations** of $f(x) = x^2$.

$f(x) = x^2$

Vertical Translations

$g(x) = x^2 - 3$

$g(x) = x^2 + 2$

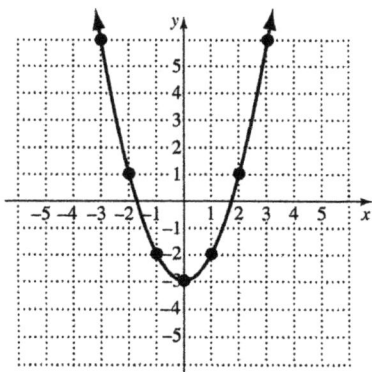

For the graph of $g(x) = x^2 - 3$, the graph of $f(x) = x^2$ is *shifted down* 3 units. We say that $g(x) = f(x) - 3$.

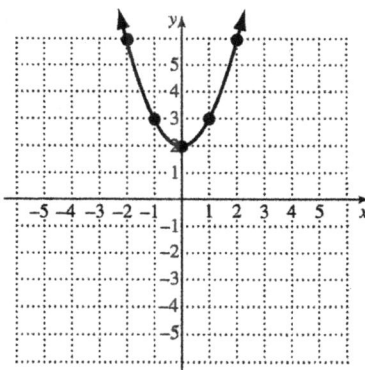

For the graph of $g(x) = x^2 + 2$, the graph of $f(x) = x^2$ is *shifted up* 2 units. We say that $g(x) = f(x) + 2$.

Horizontal Translations

$g(x) = (x-2)^2$

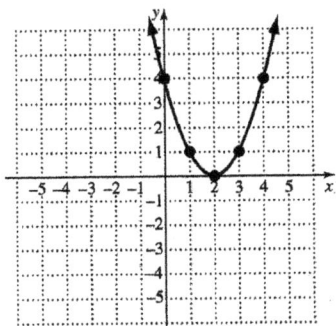

For the graph of $g(x) = (x-2)^2$, the graph of $f(x) = x^2$ is *shifted right* 2 units. We say that $g(x) = f(x-2)$.

$g(x) = (x+3)^2$

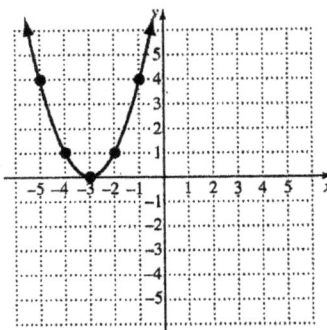

For the graph of $g(x) = (x+3)^2$, the graph of $f(x) = x^2$ is *shifted left* 3 units. We say that $g(x) = f(x+3)$.

Vertical Stretching

$g(x) = 2x^2$

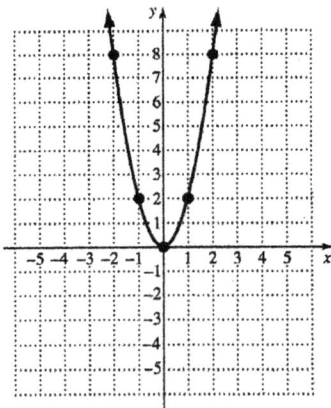

The graph of $g(x) = 2x^2$ is a *vertical stretching* of $f(x) = x^2$ by a factor of 2.

We say that $g(x) = 2f(x)$.

Vertical Shrinking

$g(x) = \frac{1}{2}x^2$

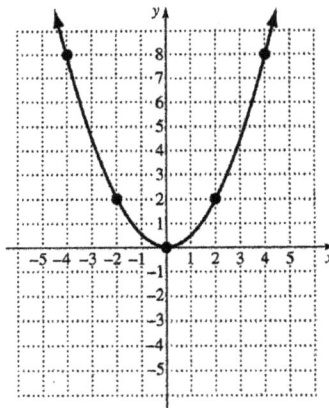

The graph of $g(x) = \frac{1}{2}x^2$ is a *vertical shrinking* of $f(x) = x^2$ by a factor of $\frac{1}{2}$.

We say that $g(x) = \frac{1}{2}f(x)$.

Horizontal Stretching

$$g(x) = \left(\frac{1}{2}x\right)^2$$

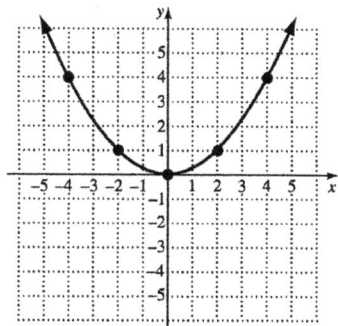

The graph of $g(x) = \left(\frac{1}{2}x\right)^2$ is

a *horizontal stretching* of $f(x) = x^2$.
The points of $g(x)$ can be found by
dividing the x-coordinates of the
points of $f(x)$ by $\frac{1}{2}$ (which is the
same as multiplying by 2). We
have $g(x) = f\left(\frac{1}{2}x\right)$.

Horizontal Shrinking

$$g(x) = (2x)^2$$

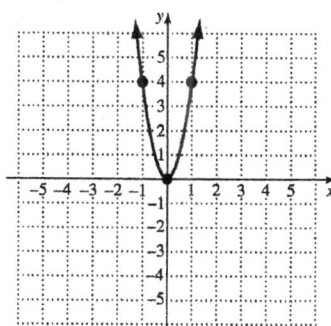

The graph of $g(x) = (2x)^2$ is

a *horizontal shrinking* of $f(x) = x^2$.
The points of $g(x)$ can be found by
dividing the x-coordinates of the
points of $f(x)$ by 2. We have

$$g(x) = f(2x).$$

Reflection Across the x-Axis

$$g(x) = -x^2$$

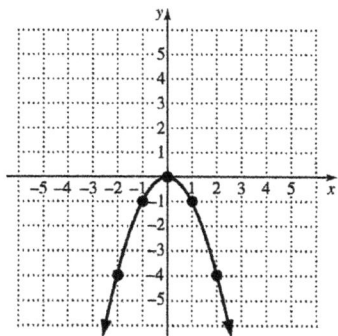

The graph of $g(x) = -x^2$ is a
reflection of the graph of $f(x)$
across the x-axis. We have

$$g(x) = -f(x).$$

Reflection Across the y-Axis

$$g(x) = (-x)^2$$

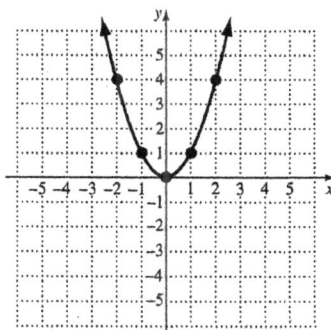

The graph of $g(x) = (-x)^2$ is a
reflection of the graph of $f(x)$
across the y-axis. Note: $(-x)^2 = x^2$,
so $g(x) = f(-x) = f(x)$. The graph
of $f(x) = x^2$ is symmetric with respect
to the y-axis; thus $f(x) = f(-x)$.

Exercises

The graph of $f(x) = |x|$ (the absolute-value function) is shown in figure (a) below. In Exercises 1-10, match the function g with one of the graphs (a) – (h) that follow. Some graphs will be used more than once.

(a)

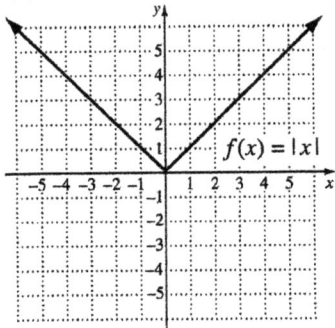

$f(x) = |x|$

(b)

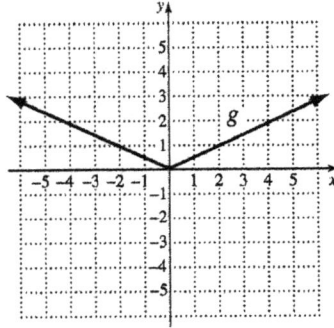

g

(c)

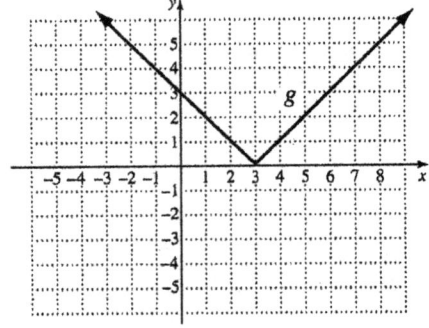

g

(d)

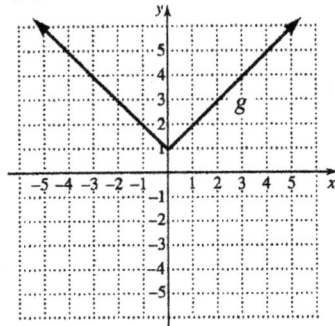

g

(e)

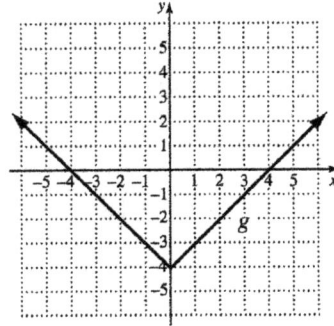

g

(f)

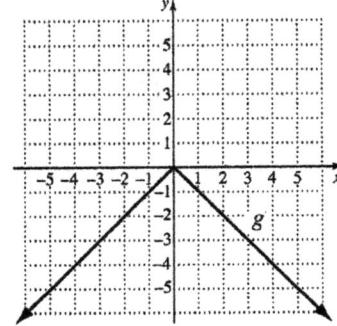

g

(g)

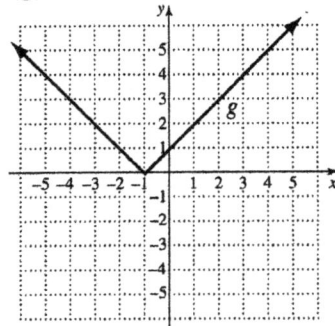

g

(h)

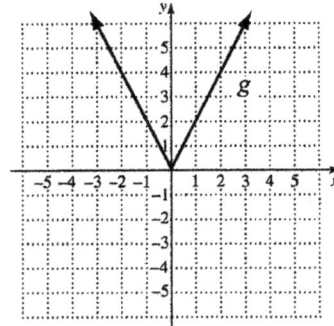

g

1. $g(x) = |x+1|$

2. $g(x) = 2|x|$

3. $g(x) = |x| + 1$

4. $g(x) = |-x|$

5. $g(x) = \dfrac{1}{2}|x|$

6. $g(x) = |x-3|$

7. $g(x) = \left|\dfrac{1}{2}x\right|$

8. $g(x) = |x| - 4$

9. $g(x) = |2x|$

10. $g(x) = -|x|$

The graph of the function *f* is shown in figure (a) below. In Exercises 11-20, match the function *g* with one of the graphs (b) – (k) that follow.

(a)

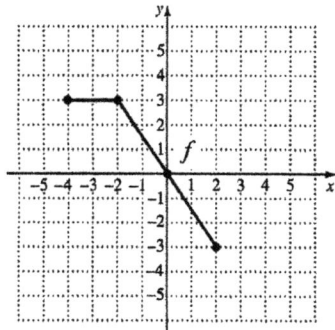

(b)

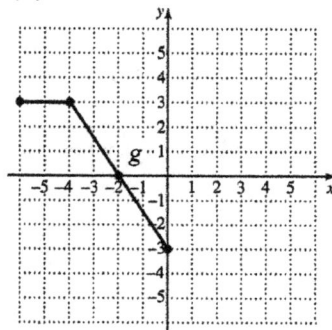

(c)

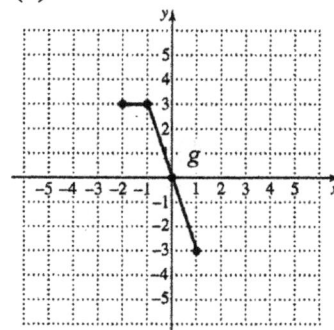

(d)

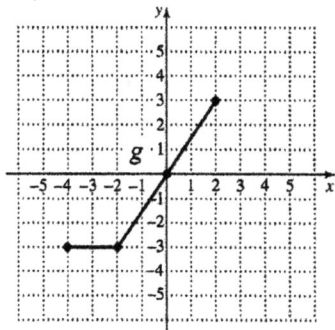

(e)

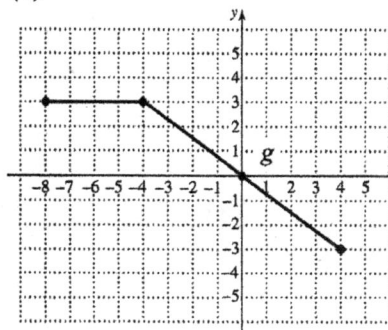

(f)

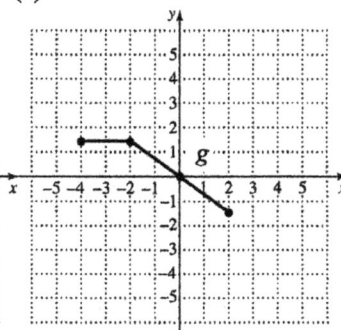

(g)

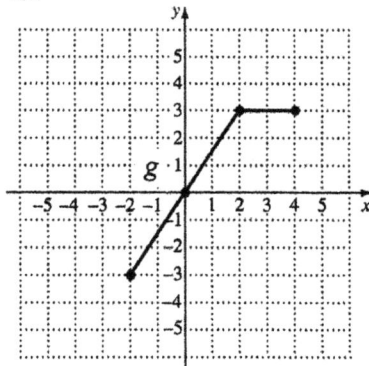

(h)

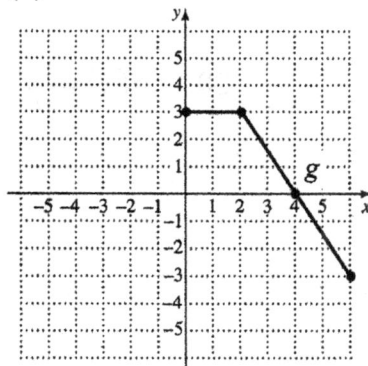

(i)

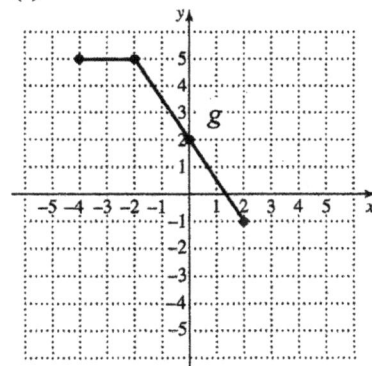

(j)

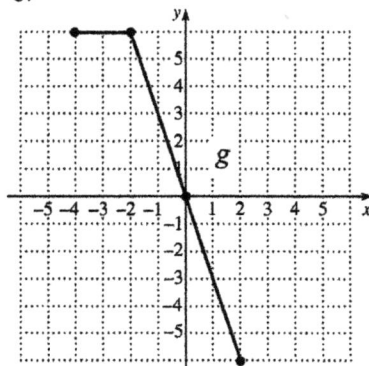

(k)

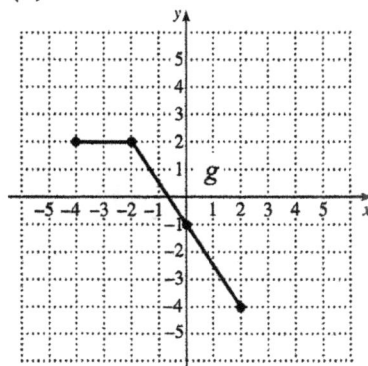

11. $g(x) = f\left(\dfrac{1}{2}x\right)$

12. $g(x) = f(-x)$

13. $g(x) = f(x-4)$

14. $g(x) = f(x) - 1$

15. $g(x) = f(2x)$

16. $g(x) = f(x) + 2$

17. $g(x) = f(x+2)$

18. $g(x) = 2f(x)$

19. $g(x) = -f(x)$

20. $g(x) = \dfrac{1}{2}f(x)$

Notes:

Interactive Preview 4: INTRODUCTION TO QUADRATIC FUNCTIONS

A **quadratic function** f is a function that can be written in the form

$$f(x) = ax^2 + bx + c, \quad a \neq 0,$$

where a, b, and c are real numbers.

Graphs of Quadratic Functions

- The graph of a quadratic function is called a **parabola**.

- The point (h, k) at which the graph turns is called the **vertex**.

- If $a > 0$ (the graph opens up), the second coordinate k of the vertex (h, k) is a **minimum value** of the function.

- If $a < 0$ (the graph opens down), the second coordinate k of the vertex (h, k) is a **maximum value** of the function.

- The **axis of symmetry**, $x = h$, is a vertical line that passes through the vertex (h, k).

- The **zeros of the function** are the first coordinates of the x-intercepts of the graph.

Example:

$$f(x) = x^2 - 2x - 8$$
$$= (x + 2)(x - 4)$$

$a = 1;$

$a > 0$, thus the graph opens up.

The zeros of $f(x)$ are -2 and 4.

The minimum value of $f(x)$ is -9.

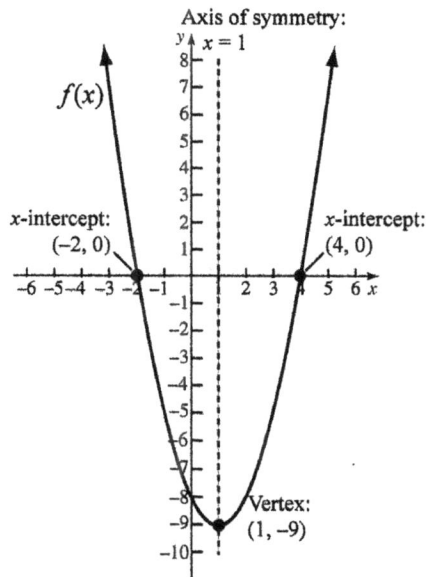

Exercises Label the vertex, the axis of symmetry, and the *x*-intercepts of the graph of the quadratic function. Then fill in the blanks in the statements below the graph.

1. $h(x) = x^2 - 8x + 15$

Axis of symmetry: _____

h(x)

x-intercept: (,)

x-intercept: (,)

Vertex: (,)

- $a = 1$; $a\ \boxed{}\ 0$, thus the graph opens _____ .
  $\underline{\text{up / down}}$

- The zeros of $h(x)$ are $\boxed{}$ and $\boxed{}$.

- The _____
  $\underline{\text{maximum / minimum}}$
  value of $h(x)$ is $\boxed{}$.

2. $f(x) = -x^2 - 4x - 3$

Vertex: (,)

x-intercept: (,)

x-intercept: (,)

f(x)

Axis of symmetry: _____

- $a = -1$; $a\ \boxed{}\ 0$, thus the graph opens _____ .
  $\underline{\text{up / down}}$

- The zeros of $f(x)$ are $\boxed{}$ and $\boxed{}$.

- The _____
  $\underline{\text{maximum / minimum}}$
  value of $f(x)$ is $\boxed{}$.

3. $g(x) = -x^2 - 2x + 3$

Vertex: (,)

g(x)

x-intercept: (,)

x-intercept: (,)

Axis of symmetry: _____

- $a = -1$; $a\ \boxed{}\ 0$, thus the graph opens _____ .
  $\underline{\text{up / down}}$

- The zeros of $g(x)$ are $\boxed{}$ and $\boxed{}$.

- The _____
  $\underline{\text{maximum / minimum}}$
  value of $g(x)$ is $\boxed{}$.

4. $f(x) = x^2 - 4x$

f(x)

x-intercept: (,)

x-intercept: (,)

Axis of symmetry: _____

Vertex: (,)

- $a = 1$; $a\ \boxed{}\ 0$, thus the graph opens _____ .
  $\underline{\text{up / down}}$

- The zeros of $f(x)$ are $\boxed{}$ and $\boxed{}$.

- The _____
  $\underline{\text{maximum / minimum}}$
  value of $f(x)$ is $\boxed{}$.

Interactive Preview 5: SOLVING RATIONAL EQUATIONS

Equations containing rational expressions are called **rational equations**. To solve a rational equation, the first step is to *clear the equation of fractions*. To do this, multiply all terms on both sides of the equation by the least common denominator (LCD) of all the rational expressions. The LCD is the least common multiple (LCM) of the denominators.

Example 1 Solve: $\dfrac{1}{4} - \dfrac{5}{6} = \dfrac{1}{a}$.

The LCM of 4, 6, and a is $12a$. Multiply on both sides by the LCD, $12a$.

$$12a\left(\frac{1}{4} - \frac{5}{6}\right) = 12a \cdot \frac{1}{a}$$

Remove parentheses.

$$12a \cdot \frac{1}{4} - 12a \cdot \frac{5}{6} = 12a \cdot \frac{1}{a}$$

Simplify.

$$3a - 10a = 12$$

Collect like terms.

$$-7a = 12$$

Divide by -7.

$$a = -\frac{12}{7}$$

The solution is $-\dfrac{12}{7}$.

Check:

$$\frac{1}{4} - \frac{5}{6} = \frac{1}{a}$$

$$\begin{array}{c|c} \dfrac{1}{4} \cdot \dfrac{3}{3} - \dfrac{5}{6} \cdot \dfrac{2}{2} & \dfrac{1}{-\dfrac{12}{7}} \\[2ex] \dfrac{3}{12} - \dfrac{10}{12} & 1 \cdot \left(-\dfrac{7}{12}\right) \\[2ex] -\dfrac{7}{12} & -\dfrac{7}{12} \quad \text{True} \end{array}$$

Example 2 Solve: $\dfrac{x-1}{15} - \dfrac{x+2}{10} = 0$.

Multiply on both sides by the LCD, 30.

$$30\left(\frac{x-1}{15} - \frac{x+2}{10}\right) = 30 \cdot 0$$

Remove parentheses.

$$30 \cdot \frac{x-1}{15} - 30 \cdot \frac{x+2}{10} = 30 \cdot 0$$

$$2(x-1) - 3(x+2) = 0$$

Simplify.

$$2x - 2 - 3x - 6 = 0$$

Collect like terms.

$$-x - 8 = 0$$

Add x.

$$-8 = x$$

The solution is -8.

Check:

$$\frac{x-1}{15} - \frac{x+2}{10} = 0$$

$$\begin{array}{c|c} \dfrac{-8-1}{15} - \dfrac{-8+2}{10} & 0 \\[2ex] -\dfrac{9}{15} - \dfrac{-6}{10} & \\[2ex] -\dfrac{3}{5} + \dfrac{3}{5} & \\[2ex] 0 & 0 \quad \text{True} \end{array}$$

Exercises Solve.

1. $\dfrac{5}{8} - \dfrac{2}{5} = \dfrac{1}{x}$

2. $\dfrac{t+3}{4} - \dfrac{t-3}{5} = 2$

Example 3   Solve: $\dfrac{7}{5y-2} = \dfrac{5}{y-1}.$

Multiply on both sides by the LCD, $(5y-2)(y-1)$.	$(5y-2)(y-1) \cdot \dfrac{7}{5y-2} = (5y-2)(y-1) \cdot \dfrac{5}{y-1}$
Simplify.	$7(y-1) = 5(5y-2)$
Remove parentheses.	$7y - 7 = 25y - 10$
Subtract $7y$ and add 10.	$3 = 18y$
Divide by 18.	$\dfrac{3}{18} = \dfrac{18y}{18}$
Simplify.	$\dfrac{1}{6} = y$

The solution is $\dfrac{1}{6}$.

Exercises Solve.

3. $\dfrac{4}{x-1} = \dfrac{3}{x+2}$

4. $\dfrac{-2}{c+6} = \dfrac{8}{3c-1}$

Example 4   Solve: $\dfrac{2t}{t-1}=\dfrac{7}{t-4}$.

Multiply on both sides by the LCD, $(t-1)(t-4)$.	$(t-1)(t-4)\cdot\dfrac{2t}{t-1}=(t-1)(t-4)\cdot\dfrac{7}{t-4}$
Simplify.	$2t(t-4)=7(t-1)$
Remove parentheses.	$2t^2-8t=7t-7$
Subtract $7t$ and add 7 to get 0 on one side.	$2t^2-15t+7=0$
Factor.	$(2t-1)(t-7)=0$
Use the principle of zero products.	$2t-1=0 \ \ or \ \ t-7=0$
Solve the two equations separately.	$2t=1 \ \ or \ \ \ \ \ t=7$
	$t=\dfrac{1}{2} \ \ or \ \ \ \ \ t=7$

Both $\dfrac{1}{2}$ and 7 check and are the solutions.

Example 5   Solve: $x+\dfrac{20}{x}=9$.

Multiply on both sides by the LCD, x.	$x\left(x+\dfrac{20}{x}\right)=x\cdot 9$
Remove parentheses.	$x\cdot x+x\cdot\dfrac{20}{x}=x\cdot 9$
Simplify.	$x^2+20=9x$
Subtract $9x$ to get 0 on one side.	$x^2-9x+20=0$
Factor.	$(x-5)(x-4)=0$
Use the principle of zero products.	$x-5=0 \ \ or \ \ x-4=0$
Solve the two equations separately.	$x=5 \ \ or \ \ \ \ \ x=4$

Checks.

$$x+\frac{20}{x}=9 \qquad\qquad x+\frac{20}{x}=9$$

$$
\begin{array}{c|c}
5+\dfrac{20}{5} & 9 \\
5+4 & \\
\hline
9 & 9 \quad \text{True}
\end{array}
\qquad
\begin{array}{c|c}
4+\dfrac{20}{4} & 9 \\
4+5 & \\
\hline
9 & 9 \quad \text{True}
\end{array}
$$

The solutions are 5 and 4.

Exercises Solve.

5. $\dfrac{3y}{5y+8} = \dfrac{2}{y-4}$

6. $z + \dfrac{22}{z} = -13$

When we multiply all terms on both sides of an equation by the LCD, the resulting equation might yield numbers that are *not* solutions of the original equation. Thus, we must always check possible solutions in the original equation.

Example 6   Solve: $\dfrac{y^2}{y-6} = \dfrac{36}{y-6}$.

Multiply on both sides by the LCD, $y-6$.
$$(y-6) \cdot \dfrac{y^2}{y-6} = (y-6) \cdot \dfrac{36}{y-6}$$

Simplify.
$$y^2 = 36$$

Use the principle of square roots.
$$y = -6 \;\; or \;\; y = 6$$

Checks.

$$\dfrac{y^2}{y-6} = \dfrac{36}{y-6}$$

$\dfrac{(-6)^2}{-6-6}$	$\dfrac{36}{-6-6}$
$\dfrac{36}{-12}$	$\dfrac{36}{-12}$

True

$$\dfrac{y^2}{y-6} = \dfrac{36}{y-6}$$

$\dfrac{(6)^2}{6-6}$	$\dfrac{36}{6-6}$
$\dfrac{36}{0}$	$\dfrac{36}{0}$

Not defined

Since division by 0 is not defined, 6 is *not* a solution. The number -6 checks, so it is the solution.

Exercises Solve.

7. $\dfrac{y^2}{y-9} = \dfrac{81}{y-9}$

8. $\dfrac{8x^2}{x+5} = \dfrac{200}{x+5}$

Interactive Preview 6: CHECKING SOLUTIONS OF RADICAL EQUATIONS

A **radical equation** has a variable in one or more radicands. For example,

$$\sqrt{2x} + 3 = 4 \quad \text{and} \quad \sqrt{x+2} - \sqrt{x+3} = 8$$

are radical equations. We use the principle of powers to solve radical equations.

> ### The Principle of Powers
> For any natural number n, if an equation $a = b$ is true, then $a^n = b^n$ is true.

For an even integer n, if an equation $a^n = b^n$ is true, it *might not* be true that $a = b$. For example, $5^2 = (-5)^2$, but $5 \neq -5$. Thus, we must check the possible solutions of radical equations containing even roots.

Example 1 Using the principle of powers, we find that the *possible* solution of $\sqrt{x} + 6 = 3$ is 9. Determine if this is the solution.

Check: Write the equation. $\quad \dfrac{\sqrt{x} + 6 = 3}{}$

Substitute 9 for x. $\quad \sqrt{9} + 6 \mid 3$

Simplify. $\quad 3 + 6$

$\quad 9 \mid 3 \quad$ False

We get a false equation, $9 = 3$, so 9 is not the solution. This equation has no solution.

Example 2 Using the principle of powers, we find that the *possible* solutions of $x = \sqrt{x+7} + 5$ are 2 and 9. Determine if these are the solutions.

For 2:

Check: Write the equation. $\quad \dfrac{x = \sqrt{x+7} + 5}{}$

Substitute 2 for x. $\quad 2 \mid \sqrt{2+7} + 5$

Simplify. $\quad \sqrt{9} + 5$

$\quad 3 + 5$

$\quad 2 \mid 8 \quad$ False

We get a false equation, $2 = 8$, so 2 is not a solution.

Next we will check 9.

Check: Write the equation. | $x = \sqrt{x+7} + 5$
    Substitute 9 for x. | $9 \;\big|\; \sqrt{9+7} + 5$
          Simplify. | $\sqrt{16} + 5$
             | $4 + 5$
             | $9 \;\big|\; 9$ True

We get a true equation, $9 = 9$, so 9 is a solution.

The solution of $x = \sqrt{x+7} + 5$ is 9.

Example 3 Using the principle of powers, we find that the *possible* solutions of $\sqrt{2m-3} + 2 = \sqrt{m+7}$ are 2 and 42. Determine if these are the solutions.

For 2:

Check: Write the equation. | $\sqrt{2m-3} + 2 = \sqrt{m+7}$
    Substitute 2 for m. | $\sqrt{2\cdot 2-3} + 2 \;\big|\; \sqrt{2+7}$
          Simplify. | $\sqrt{4-3} + 2 \;\big|\; \sqrt{9}$
          | $\sqrt{1} + 2 \;\big|\; 3$
          | $1 + 2$
          | $3 \;\big|\; 3$ True

We get a true equation, $3 = 3$, so 2 is a solution.

For 42:

Check: Write the equation. | $\sqrt{2m-3} + 2 = \sqrt{m+7}$
    Substitute 42 for m. | $\sqrt{2\cdot 42-3} + 2 \;\big|\; \sqrt{42+7}$
          Simplify. | $\sqrt{84-3} + 2 \;\big|\; \sqrt{49}$
          | $\sqrt{81} + 2 \;\big|\; 7$
          | $9 + 2$
          | $11 \;\big|\; 7$ False

We get a false equation, $11 = 7$, so 42 is not a solution.

The number 2 checks, but 42 does not, so the solution is 2.

Exercises For each equation, the possible solutions that are found using the principle of powers are given. Determine if each number is a solution.

1. $\sqrt{y} = 5$; 25

2. $\sqrt{x} + 4 = 2$; 4

3. $\sqrt{2x - 1} = 5$; 13

4. $\sqrt{6z + 3} = \sqrt{4z + 9}$; 3

5. $\sqrt{y + 2} + \sqrt{3y + 4} = 2$; −1, 7

6. $x - 7 = 2\sqrt{x+1}$; 3, 15

7. $\sqrt{2y+7} = y+2$; 1, -3

8. $\sqrt{x+2} - \sqrt{2x+2} + 1 = 0$; 7, -1

9. $\sqrt{2z-5} = 1 + \sqrt{z-3}$; 3, 7

Interactive Preview 7: SOLVING EQUATIONS AND INEQUALITIES WITH ABSOLUTE VALUE

Let's look at graphs of solution sets of equations and inequalities with absolute value.

Examples

		Equivalent Statement or Inequality	Graph of Solution Set

1. $|x| = 3$ $x = -3$ *or* $x = 3$

   The solutions are -3 and 3.

2. $|x| < 3$ $-3 < x < 3$

   The solutions are all numbers between -3 and 3. To indicate that -3 and 3 are not solutions, we use a parenthesis at -3 and at 3. In interval notation, the solution set is $(-3, 3)$.

3. $|x| \leq 3$ $-3 \leq x \leq 3$

   The solutions are all numbers from -3 to 3, including -3 and 3. To indicate that -3 and 3 are solutions, we use a bracket at -3 and at 3. In interval notation, the solution set is $[-3, 3]$.

4. $|x| > 3$ $x < -3$ *or* $x > 3$

   The solutions are all numbers less than -3 and all numbers greater than 3. Parentheses indicate that -3 and 3 are not solutions. In interval notation, the solution set is $(-\infty, -3) \cup (3, \infty)$.

5. $|x| \geq 3$ $x \leq -3$ *or* $x \geq 3$

   The solutions are all numbers less than or equal to -3 and all numbers greater than or equal to 3. Brackets indicate that -3 and 3 are solutions. In interval notation, the solution set is $(-\infty, -3] \cup [3, \infty)$.

Exercises

Graph the solution set, then state the solution. For inequalities, express the solution set in interval notation. Remember: Always use a parenthesis with $-\infty$ and ∞.

	Graph of Solution Set	Solution Set

1. $|x| = 1$ _____

2. $|z| < 2$ _____

3. $|t| \leq 5$ _____

4. $|y| > 4$

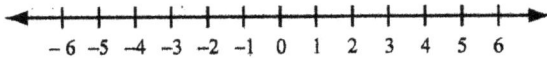

_____

5. $|y| \geq 3.5$

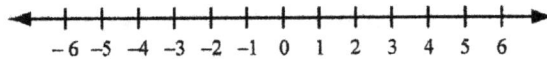

_____

The first step in solving equations and inequalities with absolute value is to write an equivalent statement or inequality that does not contain absolute value notation.

Examples Write an equivalent statement or inequality.

		Equivalent Statement or Inequality
6.	$\|w\| = 2$	$w = -2$ or $w = 2$
7.	$\|y+5\| = 14$	$y+5 = -14$ or $y+5 = 14$
8.	$\|t\| < 9$	$-9 < t < 9$
9.	$\|2s-3\| \leq 10$	$-10 \leq 2s-3 \leq 10$
10.	$\|z\| \geq 29$	$z \leq -29$ or $z \geq 29$
11.	$\|4-x\| > 8$	$4-x < -8$ or $4-x > 8$

Exercises Write an equivalent statement or inequality.

Equivalent Statement or Inequality

6. $|q| = 8$ _____

7. $|p+5| \leq 2$ _____

8. $|0.5 - x| > 6.5$ _____

9. $|7y-1| = 12$ _____

10. $\left|\dfrac{3}{4}t + 6\right| < 21$ _____

11. $|18q| \geq 9$ _____

Examples Solve and graph the solution set.

12. $|x-2|=1$

Write an equivalent statement.	$x-2=-1$ $\quad or \quad$ $x-2=1$
Add 2.	$x=1$ $\quad or \quad$ $x=3$
State the solution set.	The solutions are 1 and 3.

Graph the solution set.

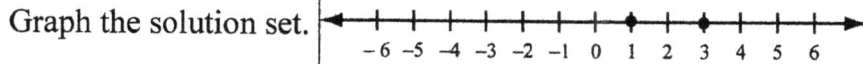

13. $|x+1|\geq 3$

Write an equivalent inequality.	$x+1\leq -3$ $\quad or \quad$ $x+1\geq 3$
Subtract 1.	$x\leq -4$ $\quad or \quad$ $x\geq 2$
State the solution set.	$(-\infty,-4]\cup[2,\infty)$

Graph the solution set.

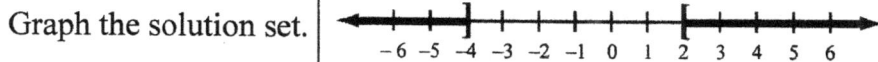

14. $|x-3|<1$

Write an equivalent inequality.	$-1<x-3<1$
Add 3.	$2<x<4$
State the solution set.	$(2,4)$

Graph the solution set.

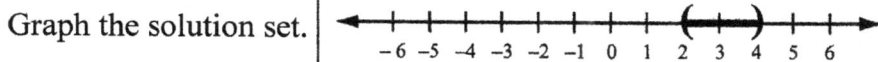

Exercises Solve and graph the solution set. In Exercises 12 and 13, fill in the blanks in key steps.

12. $|w+10|\leq 30$

$\boxed{}\leq w+10\leq 30$

$-40\leq w\leq\boxed{}$

Solution set: $\left[\boxed{},20\right]$

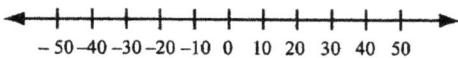

13. $|y-4|>6$

$y-4<\boxed{}$ $\quad or \quad$ $y-4>6$

$y<-2$ $\quad or \quad$ $y>\boxed{}$

Solution set: $\left(\boxed{},-2\right)\cup\left(\boxed{},\infty\right)$

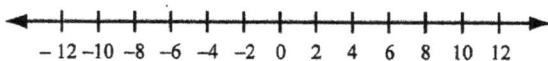

14. $|t - 2| = 3$

15. $|x - 4| < 1$

Solutions: _____

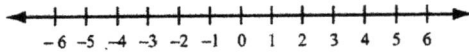

Solution set: _____

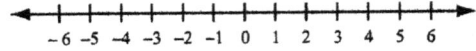

16. $|b + 5| \geq 5$

17. $\left|y + \dfrac{4}{5}\right| = \dfrac{1}{5}$

Solution set: _____

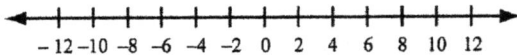

Solutions: _____

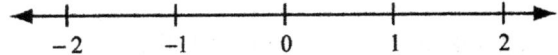

18. $|z + 1| > 1$

19. $\left|x - \dfrac{1}{2}\right| \leq \dfrac{5}{2}$

Solution set: _____

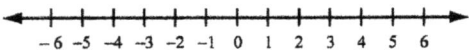

Solution set: _____

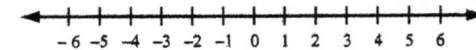

Examples Solve and graph the solution set.

15. $|2x+4|>6$

Write an equivalent inequality.	$2x+4<-6 \quad or \quad 2x+4>6$
Subtract 4.	$2x<-10 \quad or \quad 2x>2$
Divide by 2.	$x<-5 \quad or \quad x>1$
State the solution set.	$(-\infty,-5)\cup(1,\infty)$
Graph the solution set.	

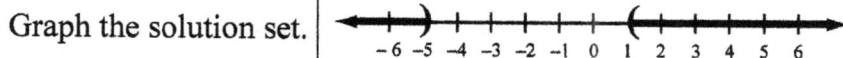

16. $|3-t|\le 2$

Write an equivalent inequality.	$-2\le 3-t\le 2$
Subtract 3.	$-5\le -t\le -1$
Multiply by -1; reverse the inequality symbols.	$5\ge t\ge 1$
Rearrange.	$1\le t\le 5$
State the solution set.	$[1,5]$
Graph the solution set.	

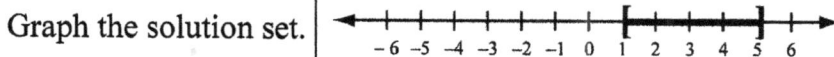

Exercises Solve and graph the solution set.

20. $|3x-6|\le 9$

21. $|2-t|>4$

Solution set: _____

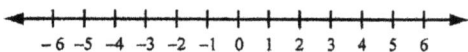

Solution set: _____

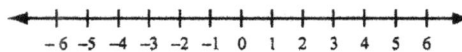

22. $|5 - t| = 15$

23. $|4x + 2| \geq 6$

Solutions: _____

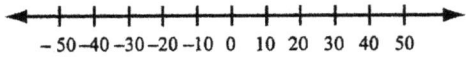

Solution set: _____

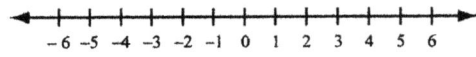

24. $\left|\dfrac{1}{4}x + \dfrac{1}{2}\right| < \dfrac{3}{2}$

25. $|30 - 10x| \leq 40$

Solution set: _____

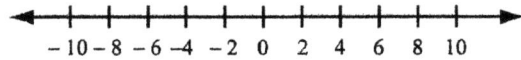

Solution set: _____

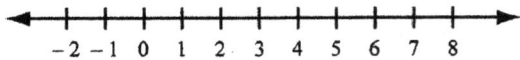

Notes:

Interactive Preview 8: ZEROS OF POLYNOMIAL FUNCTIONS

The **zeros of a function** $y = f(x)$ are also the **solutions of the equation** $f(x) = 0$, and the *real-number* zeros are the **first coordinates of the x-intercepts** of the graph of the function.

Examples

1. Linear Function

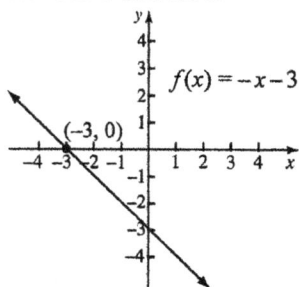

The *x-intercept* of the graph of
$$f(x) = -x - 3 \text{ is}$$
$$(-3, 0).$$
The *solution* of the equation
$$-x - 3 = 0 \text{ is}$$
$$-3.$$
The *zero* of the function
$$f(x) = -x - 3 \text{ is}$$
$$-3.$$

2. Quadratic Function

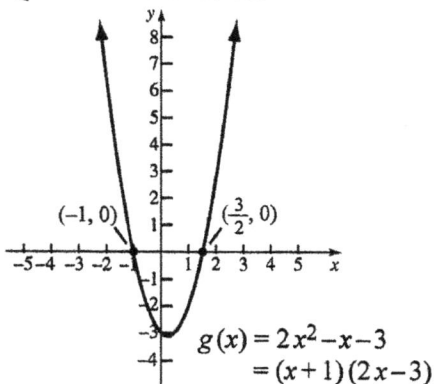

The *x-intercepts* of the graph of
$$g(x) = 2x^2 - x - 3 \text{ are}$$
$$(-1, 0) \text{ and } \left(\frac{3}{2}, 0\right).$$
The *solutions* of the equation
$$2x^2 - x - 3 = 0, \text{ or } (x+1)(2x-3) = 0, \text{ are}$$
$$-1 \text{ and } \frac{3}{2}.$$
The *zeros* of the function
$$g(x) = 2x^2 - x - 3, \text{ or } g(x) = (x+1)(2x-3), \text{ are}$$
$$-1 \text{ and } \frac{3}{2}.$$

3. Cubic Function

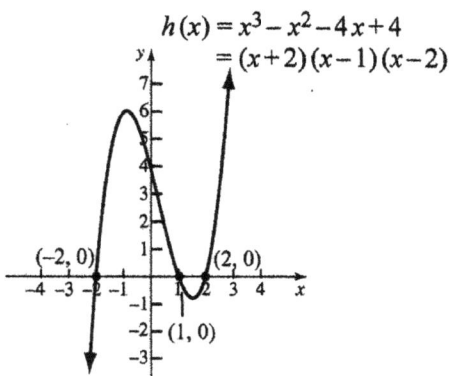

The *x-intercepts* of the graph of
$$h(x) = x^3 - x^2 - 4x + 4 \text{ are}$$
$$(-2, 0), \ (1, 0), \text{ and } (2, 0).$$
The *solutions* of the equation
$$x^3 - x^2 - 4x + 4 = 0, \text{ or } (x+2)(x-1)(x-2) = 0, \text{ are}$$
$$-2, \ 1, \text{ and } 2.$$
The *zeros* of the function
$$h(x) = x^3 - x^2 - 4x + 4, \text{ or}$$
$$h(x) = (x+2)(x-1)(x-2) \text{ are}$$
$$-2, \ 1, \text{ and } 2.$$

Exercises Label the x-intercepts of each graph. Then complete the statements with the x-intercepts of the graph of the function $f(x)$, the solutions of the equation $f(x) = 0$, and the zeros of the function $f(x)$.

1.

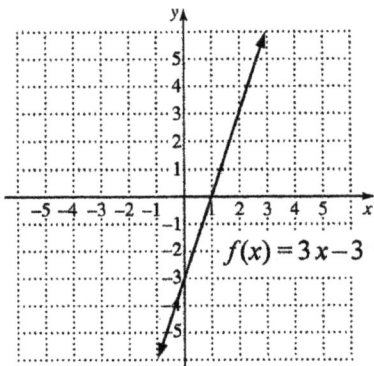

The x-intercept of the graph of
$f(x) = 3x - 3$ is _____.

The solution of the equation
$3x - 3 = 0$ is _____.

The zero of the function
$f(x) = 3x - 3$ is _____.

2.

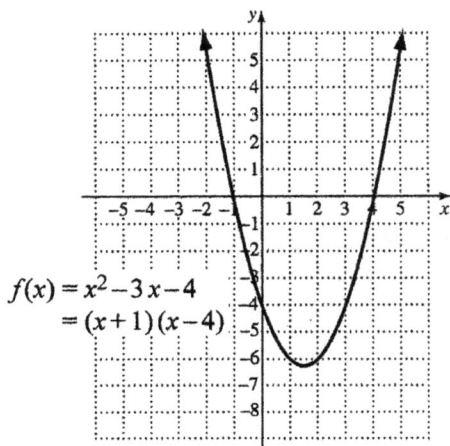

$f(x) = x^2 - 3x - 4$
$= (x+1)(x-4)$

The x-intercepts of the graph of
$f(x) = x^2 - 3x - 4$ are _____ and _____.

The solutions of the equation
$x^2 - 3x - 4 = 0$, or $(x+1)(x-4) = 0$, are
_____ and _____.

The zeros of the function
$f(x) = x^2 - 3x - 4$, or $f(x) = (x+1)(x-4)$ are
_____ and _____.

3.

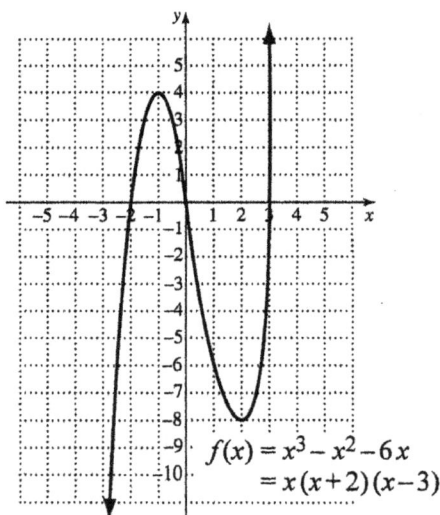

$f(x) = x^3 - x^2 - 6x$
$= x(x+2)(x-3)$

The x-intercepts of the graph of
$f(x) = x^3 - x^2 - 6x$ are _____, _____, and _____.

The solutions of the equation
$x^3 - x^2 - 6x = 0$, or $x(x+2)(x-3) = 0$, are
_____, _____, and _____.

The zeros of the function
$f(x) = x^3 - x^2 - 6x$, or $f(x) = x(x+2)(x-3)$,
are _____, _____, and _____.

Examples Using the *principle of zero products*, find the zeros of the functions in Examples 1-3.

4. $f(x) = -x - 3$ (See Example 1.)

We solve $f(x) = 0$.

$f(x) = 0$	$-x - 3 = 0$
Add x on both sides.	$-3 = x$
	The zero of $f(x)$ is -3.

5. $g(x) = 2x^2 - x - 3$ (See Example 2.)

     $= (x + 1)(2x - 3)$

We solve $g(x) = 0$.

$g(x) = 0$	$2x^2 - x - 3 = 0$
	$(x + 1)(2x - 3) = 0$
Use the principle of zero products.	$x + 1 = 0$ or $2x - 3 = 0$
Solve the equations separately.	$x = -1$ or $2x = 3$
	$x = -1$ or $x = \dfrac{3}{2}$
	The zeros of $g(x)$ are -1 and $\dfrac{3}{2}$.

6. $h(x) = x^3 - x^2 - 4x + 4$ (See Example 3.)

     $= (x + 2)(x - 1)(x - 2)$

We solve $h(x) = 0$.

$h(x) = 0$	$x^3 - x^2 - 4x + 4 = 0$
	$(x + 2)(x - 1)(x - 2) = 0$
Use the principle of zero products.	$x + 2 = 0$ or $x - 1 = 0$ or $x - 2 = 0$
Solve the equations separately.	$x = -2$ or $x = 1$ or $x = 2$
	The zeros of $h(x)$ are -2, 1, and 2.

Exercises Using the principle of zero products, find the zeros of the functions in Exercises 1-3. Show your work.

4. $f(x) = 3x - 3$ (See Exercise 1.)

5. $f(x) = x^2 - 3x - 4$ (See Exercise 2.)

       $= (x+1)(x-4)$

6. $f(x) = x^3 - x^2 - 6x$ (See Exercise 3.)

       $= x(x+2)(x-3)$

Interactive Preview 9: ASYMPTOTES OF RATIONAL FUNCTIONS

Rational Function

A rational function f is a function that is a quotient of two polynomials

$$f(x) = \frac{p(x)}{q(x)},$$

where $p(x)$ and $q(x)$ are polynomials and $q(x)$ is not the zero polynomial.

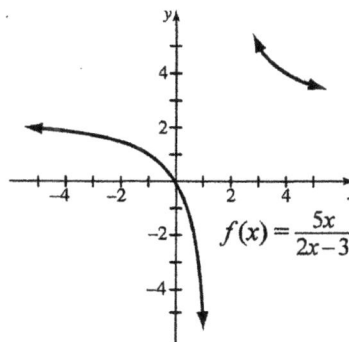

$f(x) = \dfrac{5x}{2x-3}$

When graphing rational functions, it is helpful to first sketch the asymptotes. Asymptotes can be vertical, horizontal, or oblique (slanted). In this worksheet, we will consider only rational functions with horizontal and vertical asymptotes.

Let's begin by reviewing graphs of horizontal and vertical lines. The graph of an equation of the form $x = a$ is a vertical line with x-intercept $(a, 0)$. The graph of an equation of the form $y = b$ is a horizontal line with y-intercept $(0, b)$. The graphs of $x = -3$, $y = 2$, $x = 0$, and $y = 0$ are shown below.

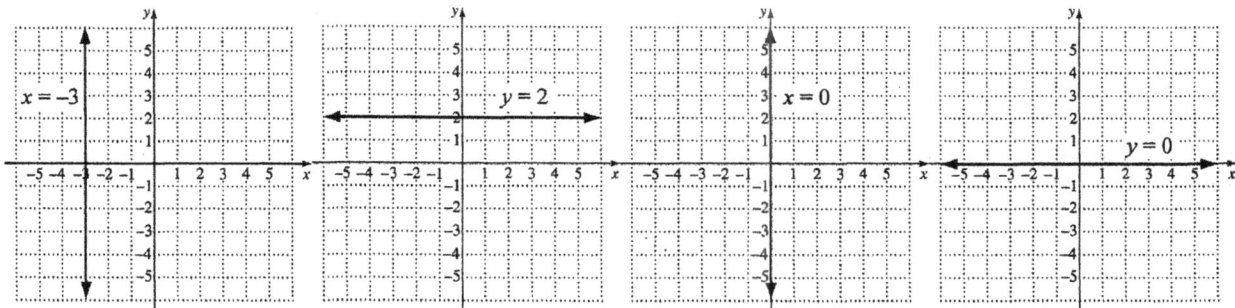

Example 1 Consider the graph of the rational function $f(x) = \dfrac{-6x-1}{3x-3}$ and observe the equations of the asymptotes.

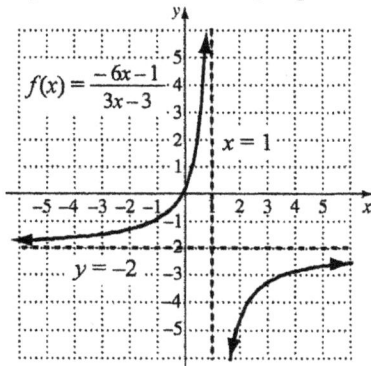

As x-values get closer to 1 from the left, the function values (y-values) approach positive infinity. Also, as the x-values get closer to 1 from the right, the function values (y-values) approach negative infinity. The vertical line $x = 1$ is the **vertical asymptote** for this curve.

As x-values approach positive infinity, function values approach -2. Likewise, as x-values approach negative infinity, function values approach -2. The horizontal line $y = -2$ is the **horizontal asymptote** for this curve.

Exercises

Label with equations all vertical and horizontal asymptotes of the graph.

1. $f(x) = \dfrac{5x-1}{5x-10}$

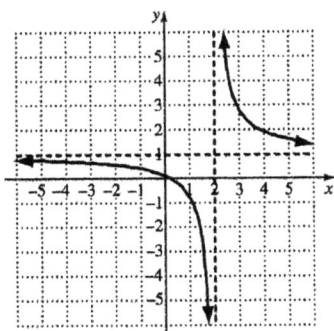

2. $f(x) = \dfrac{1}{x^2}$

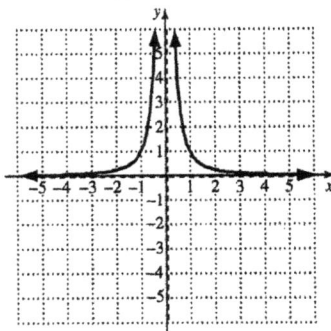

3. $f(x) = \dfrac{4}{x+2}$

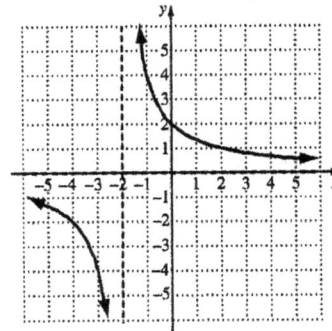

4. $f(x) = \dfrac{1+6x}{6+3x}$

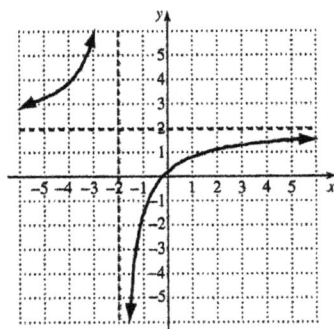

5. $f(x) = \dfrac{5}{x^2+3x}$

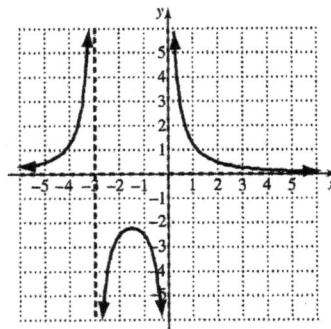

6. $f(x) = \dfrac{-2x^2+4x+3}{x^2-x-2}$

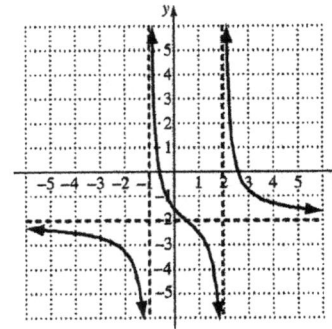

Determining Vertical Asymptotes

For a rational function $f(x) = p(x)/q(x)$, where $p(x)$ and $q(x)$ are polynomials with no common factors other than constants, if a is a zero of the denominator, then the line $x = a$ is a **vertical asymptote** for the graph of the function.

Example 2 Determine algebraically the vertical asymptotes for the graph of

$$f(x) = \frac{-6x-1}{3x-3}.$$

The numerator and the denominator do not have a common factor, so we need to determine only the zeros of the denominator $3x - 3$.

$$3x - 3 = 0$$
$$3x = 3$$
$$x = 1$$

Solving $3x - 3 = 0$, we see that 1 is a zero of the denominator. Thus, $x = 1$ is a **vertical asymptote**, as shown in the graph in Example 1.

Copyright © 2020 Pearson Education, Inc.

Exercises

Determine algebraically the vertical asymptote(s) for the graph of the function. The numerator and the denominator of each function in Exercises 7-12 do not have a common factor so we need to find only the zeros of the denominator.

7. $f(x) = \dfrac{5x-1}{5x-10}$

    Solve: $5x-10=0$.

    Zero(s) of the denominator: _____
    Vertical asymptote(s): _____
    (View the graph in Exercise 1.)

8. $f(x) = \dfrac{1}{x^2}$

    Solve: $x^2 = 0$.

    Zero(s) of the denominator: _____
    Vertical asymptote(s): _____
    (View the graph in Exercise 2.)

9. $f(x) = \dfrac{4}{x+2}$

    Solve: $x+2=0$.

    Zero(s) of the denominator: _____
    Vertical asymptote(s): _____
    (View the graph in Exercise 3.)

10. $f(x) = \dfrac{1+6x}{6+3x}$

    Solve: $6+3x=0$.

    Zero(s) of the denominator: _____
    Vertical asymptote(s): _____
    (View the graph in Exercise 4.)

11. $f(x) = \dfrac{5}{x^2+3x}$

    Solve: $x^2+3x=0$.

    Zero(s) of the denominator: _____
    Vertical asymptote(s): _____
    (View the graph in Exercise 5.)

12. $f(x) = \dfrac{-2x^2+4x+3}{x^2-x-2}$

    Solve: $x^2-x-2=0$.

    Zero(s) of the denominator: _____
    Vertical asymptote(s): _____
    (View the graph in Exercise 6.)

Determining Horizontal Asymptotes

- **Degree of Numerator = Degree of Denominator**
  When the numerator and the denominator of a rational function have the same degree, the line $y = a/b$ is the **horizontal asymptote**, where a and b are the leading coefficients of the numerator and the denominator, respectively. (For example, see the graphs in Exercises 1, 4, and 6.)

- **Degree of Numerator < Degree of Denominator**
  When the degree of the numerator of a rational function is less than the degree of the denominator, the x-axis, or $y = 0$, is the **horizontal asymptote**. (For example, see the graphs in Exercises 2, 3, and 5.)

- **Degree of Numerator > Degree of Denominator**
  When the degree of the numerator of a rational function is greater than the degree of the denominator, there is no horizontal asymptote.

Example 3 Determine algebraically the horizontal asymptote for the graph of

$$f(x) = \frac{-6x-1}{3x-3}.$$
Degree of numerator: 1
Degree of denominator: 1

The numerator and the denominator have the *same* degree. The leading coefficient of the numerator is −6. The leading coefficient of the denominator is 3. The ratio of the coefficients is −6/3, or −2. The *horizontal asymptote* is $y = -2$, as shown in the graph in Example 1.

Exercises

Determine algebraically the horizontal asymptote for the graph of the function. In Exercises 13-18, we consider only functions in which the numerator and the denominator have the same degree, or the degree of the numerator is less than the degree of the denominator.

13. $f(x) = \dfrac{5x-1}{5x-10}$

     Degree of numerator: _____
     Degree of denominator: _____
     Degree of numerator is _____ degree of denominator.
                     same as / less than
     Leading coefficient of numerator: _____
     Leading coefficient of denominator: _____
     Ratio of leading coefficients: _____
     Horizontal asymptote: _____
     (View the graph in Exercise 1.)

14. $f(x) = \dfrac{1}{x^2}$

     Degree of numerator: _____
     Degree of denominator: _____
     Degree of numerator is _____ degree of denominator.
                     same as / less than
     Horizontal asymptote: _____
     (View the graph in Exercise 2.)

15. $f(x) = \dfrac{4}{x+2}$

     Horizontal asymptote: _____
     (View the graph in Exercise 3.)

16. $f(x) = \dfrac{1+6x}{6+3x}$

     Horizontal asymptote: _____
     (View the graph in Exercise 4.)

17. $f(x) = \dfrac{5}{x^2+3x}$

     Horizontal asymptote: _____
     (View the graph in Exercise 5.)

18. $f(x) = \dfrac{-2x^2+4x+3}{x^2-x-2}$

     Horizontal asymptote: _____
     (View the graph in Exercise 6.)

Interactive Preview 10: GRAPHING INVERSE FUNCTIONS

A function is one-to-one if each output has exactly one input. The function $f(x) = 3x - 4$ is a one-to-one function. Its inverse is $f^{-1}(x) = \dfrac{x+4}{3}$. The graphs of f and f^{-1} are shown below.

x	$f(x) = 3x - 4$
0	-4
1	-1
2	2
3	5

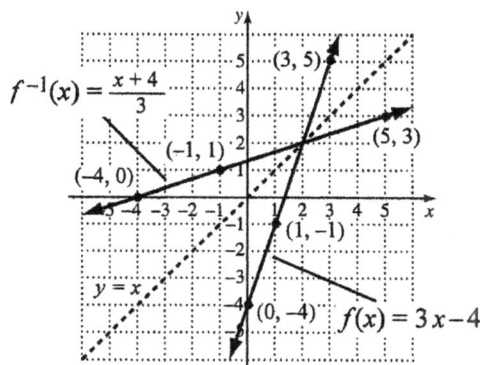

x	$f^{-1}(x) = \dfrac{x+4}{3}$
-4	0
-1	1
2	2
5	3

We can make some observations:

- The ordered-pair solutions of the inverse function, $f^{-1}(x) = \dfrac{x+4}{3}$, can be found by interchanging the first and second coordinates of each ordered-pair solution of the original function, $f(x) = 3x - 4$. For example, $(3, 5)$ is a solution of f, thus $(5, 3)$ is a solution of f^{-1}.

- The graph of $f^{-1}(x) = \dfrac{x+4}{3}$ is the reflection of the graph of $f(x) = 3x - 4$ across the line $y = x$.

One-To-One Functions and Inverses

- If a function f is one-to-one, then its inverse f^{-1} is a function.
- The domain of f is the range of f^{-1}.
- The range of f is the domain of f^{-1}.
- The solutions of f^{-1} can be found from the solutions of f by interchanging the first and second coordinates of each ordered pair.
- The graph of f^{-1} is a reflection of the graph of f across the line $y = x$.

Example 1 Graph the one-to-one function $g(x) = -2x + 3$ and its inverse $g^{-1}(x) = \dfrac{3-x}{2}$ on the same set of axes.

Begin by listing a few solutions of $g(x)$ in a table. Then plot the points and draw the graph.

x	$g(x) = -2x + 3$
-1	5
0	3
2	-1
3	-3

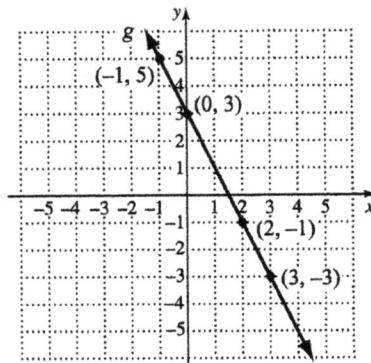

Next, create a table of solutions of $g^{-1}(x)$ by *interchanging* the x and y-values in the table for $g(x)$. Then plot the points and draw the graph of $g^{-1}(x)$ on the same set of axes as the graph of $g(x)$.

x	$g^{-1}(x) = \dfrac{3-x}{2}$
5	-1
3	0
-1	2
-3	3

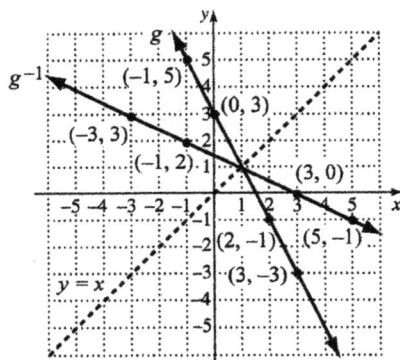

The graph of $g^{-1}(x)$ is a reflection of the graph of $g(x)$ across the line $y = x$.

When the inverse of a function is not a function, the domain of the function can be restricted to allow the inverse to be a function. Consider the graphs of $h(x) = x^2 + 1$ and its inverse $y = \pm\sqrt{x-1}$. The function $h(x) = x^2 + 1$ is not one-to-one because some outputs correspond to more than one input. Thus its inverse $y = \pm\sqrt{x-1}$ is *not* a function. However, if we restrict the domain of $h(x) = x^2 + 1$ to nonnegative numbers, $x \geq 0$, then its inverse is a function. (See Example 2.)

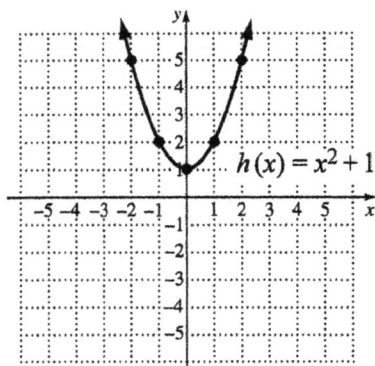

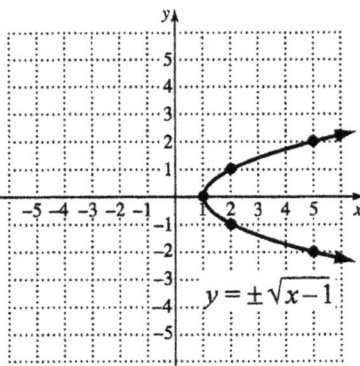

Example 2 Graph the one-to-one function $h(x) = x^2 + 1$, $x \geq 0$ and its inverse $h^{-1}(x) = \sqrt{x-1}$ on the same set of axes.

We first list solutions of $h(x) = x^2 + 1$ with the restricted domain $x \geq 0$. Then we list solutions for $h^{-1}(x)$ by interchanging the x and y-values in the table for $h(x)$ and we graph $h(x)$ and $h^{-1}(x)$ on the same set of axes.

x	$h(x) = x^2 + 1$, $x \geq 0$
0	1
1	2
2	5

x	$h^{-1}(x) = \sqrt{x-1}$
1	0
2	1
5	2

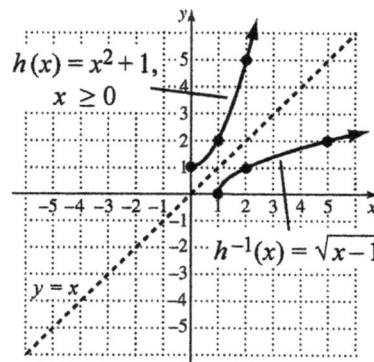

The graph of $h^{-1}(x)$ is a reflection of the graph of $h(x)$, $x \geq 0$, across the line $y = x$.

Exercises Complete tables of solutions for the one-to-one function and its inverse. Then graph both functions on the same set of axes and observe the reflection across the line $y = x$.

1. $f(x) = 2x + 4$; $f^{-1}(x) = \dfrac{x-4}{2}$

x	$f(x) = 2x + 4$
-4	
-3	
-1	
0	

x	$f^{-1}(x) = \dfrac{x-4}{2}$

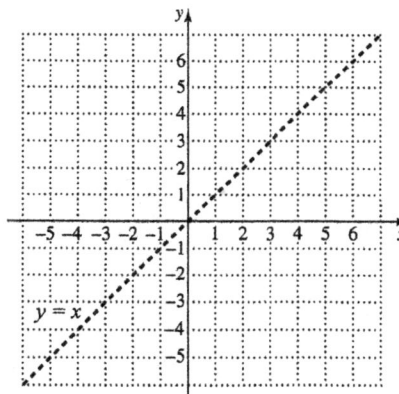

2. $h(x) = 2 - 3x$; $h^{-1}(x) = \dfrac{2-x}{3}$

x	$h(x) = 2 - 3x$
-1	
0	
1	
2	

x	$h^{-1}(x) = \dfrac{2-x}{3}$

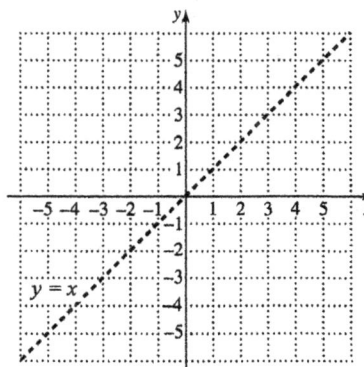

3. $g(x) = x^3 - 3$; $g^{-1}(x) = \sqrt[3]{x+3}$

x	$g(x) = x^3 - 3$
-2	
-1	
0	
1	
2	

x	$g^{-1}(x) = \sqrt[3]{x+3}$

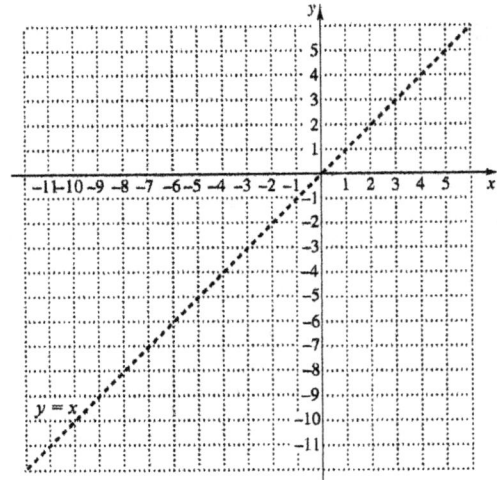

4. $f(x) = x^2 + 4,\ x \geq 0$; $f^{-1}(x) = \sqrt{x-4}$

x	$f(x) = x^2 + 4,\ x \geq 0$
0	
1	
2	
3	

x	$f^{-1}(x) = \sqrt{x-4}$

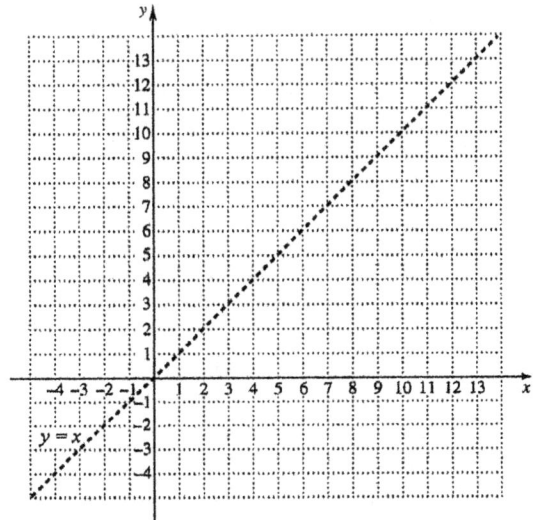

5. $h(x) = x^2 - 3,\ x \geq 0$; $h^{-1}(x) = \sqrt{x+3}$

x	$h(x) = x^2 - 3,\ x \geq 0$
0	
1	
2	
3	

x	$h^{-1}(x) = \sqrt{x+3}$

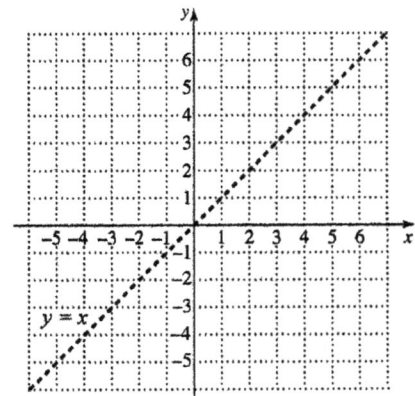

Interactive Preview 11: SOLVING EXPONENTIAL EQUATIONS

Equations with variables in the exponents, such as

$$7^x = 49, \quad 3^{4x-15} = 243, \quad 6^{y_1} = 13, \text{ and } 10e^{0.1t} = 50,$$

are called **exponential equations**.

We use the following property to solve exponential equations.

Base-Exponent Property

For any $a > 0$, $a \neq 1$,

  if $a^x = a^y$, then $x = y$, and

  if $x = y$, then $a^x = a^y$.

Example 1 Solve: $\quad 7^x = 49$.

Note that $49 = 7^2$. We can write each side as a power of 7.	$7^x = 7^2$	Check:
		$\dfrac{7^x = 49}{7^2 \mid 49}$
Use the base-exponent property.	$x = 2$	$49 \mid 49$ True

The solution is 2.

Example 2 Solve: $\quad 3^{4x-15} = 243$.

Note that $243 = 3^5$. We can write each side as a power of 3.	$3^{4x-15} = 3^5$	Check:
		$\dfrac{3^{4x-15} = 243}{3^{4 \cdot 5 - 15} \mid 243}$
Use the base-exponent property.	$4x - 15 = 5$	3^{20-15}
Add 15.	$4x = 20$	3^5
Divide by 4.	$x = 5$	$243 \mid 243$ True

The solution is 5.

Another property that is used when solving exponential equations is the property of logarithmic equality.

Property of Logarithmic Equality

For any $M > 0$, $N > 0$, $a > 0$, and $a \neq 1$,

  if $\log_a M = \log_a N$, then $M = N$, and

  if $M = N$, the $\log_a M = \log_a N$.

Example 3 Solve: $\qquad 6^x = 13.$

Take the common logarithm on both sides.	$\log 6^x = \log 13$
Use the power rule: $\log_a M^p = p \log_a M.$	$x \log 6 = \log 13$
Divide by $\log 6.$	$x = \dfrac{\log 13}{\log 6}$
Approximate using a calculator.	$x \approx 1.4315$

Check:

$$\dfrac{6^x = 13}{6^{1.4315} \mid 13}$$
$$\approx 13 \mid 13 \quad \text{True}$$

The solution is about $1.4315.$

Example 4 Solve: $\qquad e^x = 8.$

Take the natural logarithm on both sides.	$\ln e^x = \ln 8$
Use the power rule.	$x \cdot \ln e = \ln 8$
Substitute 1 for $\ln e$: $\ln e = 1.$	$x \cdot 1 = \ln 8$
	$x = \ln 8$
Approximate using a calculator.	$x \approx 2.0794$

Check:

$$\dfrac{e^x = 8}{e^{2.0794} \mid 8}$$
$$\approx 8 \mid 8 \quad \text{True}$$

The solution is about $2.0794.$

Example 5 Solve: $\qquad 4e^{3x} = 720.$

Divide by 4.	$e^{3x} = 180$
Take the natural logarithm on both sides.	$\ln e^{3x} = \ln 180$
Use the power rule.	$3x \ln e = \ln 180$
$\ln e = 1.$	$3x = \ln 180$
Divide by 3.	$x = \dfrac{\ln 180}{3}$
Approximate using a calculator.	$x \approx 1.7310$

Check:

$$\dfrac{4e^{3x} = 720}{4e^{3(1.7310)} \mid 720}$$
$$\approx 720 \mid 720 \quad \text{True}$$

The solution is about $1.7310.$

Exercises

In Exercises 1-4, describe the calculation in each step of the solution. (See Examples 1-4.)

1. Solve: $\qquad$ $5^x = 625$. (See Example 1.)

   _____ | $5^x = 5^4$
   _____ | $x = 4$
   | The solution is 4.

2. Solve: $\qquad$ $2^{3x+1} = 128$. (See Example 2.)

   _____ | $2^{3x+1} = 2^7$
   _____ | $3x + 1 = 7$
   _____ | $3x = 6$
   _____ | $x = 2$
   | The solution is 2.

3. Solve: $\qquad$ $8^x = 25$. (See Example 3.)

   _____ | $\log 8^x = \log 25$
   _____ | $x \log 8 = \log 25$
   _____ | $x = \dfrac{\log 25}{\log 8}$
   _____ | $x \approx 1.5480$
   | The solution is about 1.5480.

4. Solve: $\qquad$ $e^x = 350$. (See Example 4.)

   _____ | $\ln e^x = \ln 350$
   _____ | $x \ln e = \ln 350$
   _____ | $x = \ln 350$
   _____ | $x \approx 5.8579$
   | The solution is about 5.8579.

Solve. Approximate the answer to 4 decimal places.

5. $10^x = 1,000,000$

6. $3^{5x} = 243$

7. $e^x = 12$

8. $5^{3x} = 625$

9. $9^x = 2$

10. $e^{5x} = 120$

11. $2^x = 256$

12. $4^{5x+2} = 64$

13. $6^{4x-3} = 216$

14. $10^x = 43$

15. $6e^{3x} = 78$

16. $4^x = 17$

17. $4^{0.5x} = 16$

18. $e^x = 10,000$

19. $3^{x-10} = 729$

20. $15,000e^{0.02x} = 60,000$

21. $e^{-x} = 0.8$

22. $3^x = \dfrac{1}{27}$

Interactive Preview 12: SOLVING LOGARITHMIC EQUATIONS

Equations containing variables in logarithmic expressions, such as

$$\log_2 x = -5, \quad \log_4(7-3x) = 2,$$

$$\log_3 x + \log_3(x+8) = 2, \text{ and } \log(3x+2) - \log(3x-1) = 1,$$

are called **logarithmic equations**.

To solve logarithmic equations algebraically, we first try to obtain a single logarithmic expression on one side and then write an equivalent exponential equation.

Example 1 Solve: $\log_2 x = -5$.

Convert to an exponential equation.	$2^{-5} = x$
Write with a positive exponent.	$\dfrac{1}{2^5} = x$
	$\dfrac{1}{32} = x$

The solution is $\dfrac{1}{32}$.

Check:

$$\log_2 x = -5$$

$$\log_2\left(\dfrac{1}{32}\right) \;\Big|\; -5$$

$$\log_2 2^{-5}$$

$$-5 \;\Big|\; -5 \quad \text{True}$$

Example 2 Solve: $\log_4(7-3x) = 2$.

Convert to an exponential equation.	$4^2 = 7 - 3x$
	$16 = 7 - 3x$
Subtract 7.	$9 = -3x$
Divide by -3.	$-3 = x$

The solution is -3.

Check:

$$\log_4(7-3x) = 2$$

$$\log_4[7-3(-3)] \;\Big|\; 2$$

$$\log_4 16$$

$$2 \;\Big|\; 2 \quad \text{True}$$

Example 3 Solve: $\log x = 0$.

A common logarithm, the base is 10.	$\log_{10} x = 0$
Convert to an exponential equation.	$10^0 = x$
$a^0 = 1,\; a \neq 0$	$1 = x$

The solution is 1.

Example 4 Solve: $\log_6 x = 1$.

Convert to an exponential equation.	$6^1 = x$
$a^1 = a$	$6 = x$

The solution is 6.

> $\log_a 1 = 0,$
> for any logarithmic base a.

> $\log_a a = 1,$
> for any logarithmic base a.

Exercises Solve.

1. $\log_3 x = 5$ 2. $\log_8 x = 0$ 3. $\log_5 (18 - x) = 3$ 4. $\log_2 (3x - 7) = 4$

5. $\log_{100} x = 1$ 6. $\log(3x + 28) = 2$ 7. $\log x = -3$ 8. $\log_7 \left(\dfrac{1}{2} x + 3 \right) = 2$

Example 5 Solve: $\log_4 (5x + 12) - \log_4 (2x - 6) = 2$.

Use the quotient rule.	$\log_4 \dfrac{5x + 12}{2x - 6} = 2$
Convert to an exponential equation.	$4^2 = \dfrac{5x + 12}{2x - 6}$
	$16 = \dfrac{5x + 12}{2x - 6}$
Multiply on both sides by the LCD, $2x - 6$.	$(2x - 6) \cdot 16 = (2x - 6) \cdot \dfrac{5x + 12}{2x - 6}$
Simplify.	$32x - 96 = 5x + 12$
Subtract $5x$ and add 96.	$27x = 108$
Divide by 27.	$x = 4$
Check.	

$$\log_4 (5x + 12) - \log_4 (2x - 6) = 2$$

$$\begin{array}{l|l} \log_4 (5 \cdot 4 + 12) - \log_4 (2 \cdot 4 - 6) & 2 \\ \log_4 32 - \log_4 2 & \\ \log_4 \dfrac{32}{2} & \\ \log_4 16 & \\ 2 & 2 \quad \text{True} \end{array}$$

The solution is 4.

Exercises Solve.

9. $\log_6 x - \log_6 (x+10) = -1$

10. $\log (x+7) - \log (x-2) = 1$

11. $\log_3 (x-1) - \log_3 (3x+4) = -2$

12. $\log_5 (3x+5) - \log_5 x = 2$

Example 6 Solve: $\log_3 x + \log_3 (x+8) = 2$.

Use the product rule.	$\log_3 \left[x(x+8) \right] = 2$
Convert to an exponential equation.	$x(x+8) = 3^2$
	$x^2 + 8x = 9$
	$x^2 + 8x - 9 = 0$
Factor.	$(x+9)(x-1) = 0$
Use the principle of zero products.	$x+9 = 0 \quad or \quad x-1 = 0$
	$x = -9 \quad or \quad x = 1$
	(Checks for -9 and 1 are on the following page.)

Example 6 (continued)

Check for -9.

$$\log_3 x + \log_3 (x+8) = 2$$

$$\frac{}{\log_3 (-9) + \log_3 (-9+8)} \Big| 2$$

The number -9 is **not** a solution because negative numbers do not have real-number logarithms.

Check for 1.

$$\log_3 x + \log_3 (x+8) = 2$$

$$
\begin{array}{c|c}
\log_3 1 + \log_3 (1+8) & 2 \\
\log_3 1 + \log_3 9 & \\
0 + 2 & \\
2 & 2 \quad \text{True}
\end{array}
$$

The solution is 1.

Exercises Solve.

13. $\log_{12} (x-5) + \log_{12} (x+5) = 2$

14. $\log_8 x + \log_8 (x-7) = 1$

15. $\log (x-21) + \log x = 2$

16. $\log_3 (x+1) + \log_3 (x-1) = 3$

Interactive Preview 13: USING AN INVERSE MATRIX TO SOLVE A SYSTEM OF EQUATIONS

We begin by recalling the multiplicative inverse property and the definitions of an identity matrix and the inverse of a matrix.

Multiplicative Inverse Property

For every nonzero real number a, there is a **multiplicative inverse**, $\dfrac{1}{a}$, or a^{-1}, such that

$$a \cdot \frac{1}{a} = \frac{1}{a} \cdot a = 1, \text{ or } a \cdot a^{-1} = a^{-1} \cdot a = 1.$$

Example: 6 and $\dfrac{1}{6}$ are multiplicative inverses.

$$6 \cdot \frac{1}{6} = \frac{1}{6} \cdot 6 = 1, \text{ or } 6 \cdot 6^{-1} = 6^{-1} \cdot 6 = 1.$$

Identity Matrix

For any positive integer n, the $n \times n$ **identity matrix**, $\mathbf{I}$, is an $n \times n$ matrix with 1's on the main diagonal and 0's elsewhere.

Examples: $\mathbf{I} = \begin{bmatrix} 1 & 0 \\ 0 & 1 \end{bmatrix}$
2×2 identity matrix

and $\mathbf{I} = \begin{bmatrix} 1 & 0 & 0 \\ 0 & 1 & 0 \\ 0 & 0 & 1 \end{bmatrix}$
3×3 identity matrix

$\mathbf{A} \times \mathbf{I} = \mathbf{I} \times \mathbf{A} = \mathbf{A}$ for any $n \times n$ matrix $\mathbf{A}$.

Example: $\begin{bmatrix} 2 & -9 \\ -4 & 8 \end{bmatrix} \cdot \begin{bmatrix} 1 & 0 \\ 0 & 1 \end{bmatrix} = \begin{bmatrix} 1 & 0 \\ 0 & 1 \end{bmatrix} \cdot \begin{bmatrix} 2 & -9 \\ -4 & 8 \end{bmatrix} = \begin{bmatrix} 2 & -9 \\ -4 & 8 \end{bmatrix}$

$\qquad\quad \mathbf{A} \quad \cdot \quad \mathbf{I} \ = \ \mathbf{I} \quad \cdot \quad \mathbf{A} \quad = \quad \mathbf{A}$

Inverse of a Matrix

For an $n \times n$ matrix $\mathbf{A}$, if there is a matrix $\mathbf{A}^{-1}$ for which $\mathbf{A}^{-1} \cdot \mathbf{A} = \mathbf{A} \cdot \mathbf{A}^{-1} = \mathbf{I}$, then $\mathbf{A}^{-1}$ is the inverse of $\mathbf{A}$.

Example: $\begin{bmatrix} 8 & -5 \\ -3 & 2 \end{bmatrix}$ is the inverse of $\begin{bmatrix} 2 & 5 \\ 3 & 8 \end{bmatrix}$ because

$$\begin{bmatrix} 8 & -5 \\ -3 & 2 \end{bmatrix}\begin{bmatrix} 2 & 5 \\ 3 & 8 \end{bmatrix} = \begin{bmatrix} 2 & 5 \\ 3 & 8 \end{bmatrix}\begin{bmatrix} 8 & -5 \\ -3 & 2 \end{bmatrix} = \begin{bmatrix} 1 & 0 \\ 0 & 1 \end{bmatrix}$$

$\qquad \mathbf{A}^{-1} \ \cdot \ \mathbf{A} \qquad \mathbf{A} \ \cdot \ \mathbf{A}^{-1} \ = \ \mathbf{I}$

Let's compare solving an equation using the multiplicative inverse property with solving a system of equations using an inverse matrix and an identity matrix.

Example 1   Solve: $\frac{2}{5}x = -8$.

Multiply on both sides by $\frac{5}{2}$, the multiplicative inverse of $\frac{2}{5}$.

$$\frac{5}{2} \cdot \frac{2}{5}x = \frac{5}{2}(-8)$$

Use the multiplicative inverse property: $\frac{5}{2} \cdot \frac{2}{5} = 1$.

$$1 \cdot x = -\frac{40}{2}$$

Simplify.

$$x = -20$$

The solution is -20.

We can write a system of n linear equations in n variables as a matrix equation $\mathbf{AX} = \mathbf{B}$. If $\mathbf{A}$ has an inverse, then the system of equations has a unique solution that can be found by solving for $\mathbf{X}$.

Example 2   Solve: $2x + 5y = -1$,   The inverse of $\begin{bmatrix} 2 & 5 \\ 3 & 8 \end{bmatrix}$ is $\begin{bmatrix} 8 & -5 \\ -3 & 2 \end{bmatrix}$.
$3x + 8y = 2$.

Write an equivalent matrix equation.

$$\begin{bmatrix} 2 & 5 \\ 3 & 8 \end{bmatrix} \cdot \begin{bmatrix} x \\ y \end{bmatrix} = \begin{bmatrix} -1 \\ 2 \end{bmatrix}$$

Multiply on the left on both sides by the inverse of $\begin{bmatrix} 2 & 5 \\ 3 & 8 \end{bmatrix}$.

$$\begin{bmatrix} 8 & -5 \\ -3 & 2 \end{bmatrix} \cdot \begin{bmatrix} 2 & 5 \\ 3 & 8 \end{bmatrix} \cdot \begin{bmatrix} x \\ y \end{bmatrix} = \begin{bmatrix} 8 & -5 \\ -3 & 2 \end{bmatrix} \cdot \begin{bmatrix} -1 \\ 2 \end{bmatrix}$$

Use $\mathbf{A}^{-1} \cdot \mathbf{A} = \mathbf{I}$ on the left side and multiply on the right side.

$$\begin{bmatrix} 1 & 0 \\ 0 & 1 \end{bmatrix} \cdot \begin{bmatrix} x \\ y \end{bmatrix} = \begin{bmatrix} -18 \\ 7 \end{bmatrix}$$

Use $\mathbf{I} \cdot \mathbf{A} = \mathbf{A}$.

$$\begin{bmatrix} x \\ y \end{bmatrix} = \begin{bmatrix} -18 \\ 7 \end{bmatrix}$$

Check.

$$\begin{array}{c|c} 2x + 5y = -1 \\ \hline 2(-18) + 5 \cdot 7 & -1 \end{array}$$

Substitute -18 for x and 7 for y.

$$\begin{array}{r|ll} -36 + 35 & & \\ -1 & -1 & \text{True} \end{array}$$

$$\begin{array}{c|c} 3x + 8y = 2 \\ \hline 3(-18) + 8 \cdot 7 & 2 \end{array}$$

$$\begin{array}{r|ll} -54 + 56 & & \\ 2 & 2 & \text{True} \end{array}$$

The solution is $(-18, 7)$.

Exercises In Exercises 1-4, fill in the blanks at key steps in the problem-solving process.

1. Solve: $\dfrac{4}{9}y = -16.$

$$\boxed{} \cdot \dfrac{4}{9}y = \boxed{} \cdot (-16)$$

$$1 \cdot y = -\dfrac{\boxed{}}{4}$$

$$y = \boxed{}$$

2. Solve: $-\dfrac{5}{2}t = -\dfrac{7}{10}.$

$$\boxed{} \cdot \left(-\dfrac{5}{2}t\right) = \boxed{} \cdot \left(-\dfrac{7}{10}\right)$$

$$\boxed{} \cdot t = \dfrac{14}{\boxed{}}$$

$$t = \dfrac{\boxed{}}{25}$$

3. Solve: $-3x + 4y = -9,$
   $5x - 7y = 16.$

The inverse of $\begin{bmatrix} -3 & 4 \\ 5 & -7 \end{bmatrix}$ is $\begin{bmatrix} -7 & -4 \\ -5 & -3 \end{bmatrix}.$

Write an equivalent matrix equation, **AX = B**.

$$\begin{bmatrix} & \\ & \end{bmatrix} \cdot \begin{bmatrix} x \\ y \end{bmatrix} = \begin{bmatrix} \\ \end{bmatrix}$$

Multiply on both sides by the inverse of **A**.

$$\begin{bmatrix} & \\ & \end{bmatrix} \begin{bmatrix} -3 & 4 \\ 5 & -7 \end{bmatrix} \cdot \begin{bmatrix} x \\ y \end{bmatrix} = \begin{bmatrix} & \\ & \end{bmatrix} \cdot \begin{bmatrix} -9 \\ 16 \end{bmatrix}$$

Simplify.

$$\begin{bmatrix} & \\ & \end{bmatrix} \cdot \begin{bmatrix} x \\ y \end{bmatrix} = \begin{bmatrix} \\ \end{bmatrix}$$

Use **I · A = A**.

$$\begin{bmatrix} \\ \end{bmatrix} = \begin{bmatrix} \\ \end{bmatrix}$$

Check:

$-3x + 4y = -9$		$5x - 7y = 16$	
$-3 \cdot \boxed{} + 4 \cdot \boxed{}$	-9	$5 \cdot \boxed{} - 7 \cdot \boxed{}$	16
$\boxed{} + \boxed{}$		$\boxed{} - \boxed{}$	
$\boxed{}$	-9	$\boxed{}$	16

The solution is $\left(\boxed{}, \boxed{}\right).$

4. Solve: $5a - 4b = -6,$
$7a - 6b = -10.$

The inverse of $\begin{bmatrix} 5 & -4 \\ 7 & -6 \end{bmatrix}$ is $\begin{bmatrix} 3 & -2 \\ \dfrac{7}{2} & -\dfrac{5}{2} \end{bmatrix}$.

Write an equivalent matrix equation, $\mathbf{AX} = \mathbf{B}$.

$$\begin{bmatrix} 5 & -4 \\ 7 & -6 \end{bmatrix} \cdot \begin{bmatrix} \end{bmatrix} = \begin{bmatrix} \end{bmatrix}$$

Multiply by the inverse of $\mathbf{A}$.

$$\begin{bmatrix} \end{bmatrix} \cdot \begin{bmatrix} 5 & -4 \\ 7 & -6 \end{bmatrix} \cdot \begin{bmatrix} \end{bmatrix} = \begin{bmatrix} \end{bmatrix} \cdot \begin{bmatrix} \end{bmatrix}$$

Simplify.

$$\begin{bmatrix} \end{bmatrix} \begin{bmatrix} a \\ b \end{bmatrix} = \begin{bmatrix} \end{bmatrix}$$

Use $\mathbf{I} \cdot \mathbf{A} = \mathbf{A}$.

$$\begin{bmatrix} a \\ b \end{bmatrix} = \begin{bmatrix} \end{bmatrix}$$

The solution is $\left(\boxed{}, \boxed{} \right)$.

5. Solve: $-3x = \dfrac{6}{11}.$

6. Solve: $\dfrac{3}{4}y = -450.$

7. Solve: $2s - 3t = 16,$
$s - t = 7.$

The inverse of $\begin{bmatrix} 2 & -3 \\ 1 & -1 \end{bmatrix}$ is $\begin{bmatrix} -1 & 3 \\ -1 & 2 \end{bmatrix}$.

8. Solve: $2x + 3y = 1,$
$16x + 25y = 5.$

The inverse of $\begin{bmatrix} 2 & 3 \\ 16 & 25 \end{bmatrix}$ is $\begin{bmatrix} \dfrac{25}{2} & -\dfrac{3}{2} \\ -8 & 1 \end{bmatrix}$.

Interactive Preview 14: CLASSIFYING EQUATIONS OF CONIC SECTIONS

Conic sections can be described algebraically using second-degree equations of the form $Ax^2 + Bxy + Cy^2 + Dx + Ey + F = 0$. Circles, ellipses, hyperboles, and parabolas are conic sections. In this Preview, the circles, ellipses, and hyperbolas we consider are centered at the origin.

Circles:

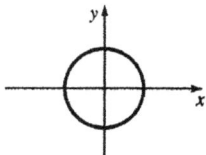

- Can be written in the form $x^2 + y^2 = r^2$, $r > 0$.
- Both variables, x and y, are squared.
- The squared terms, x^2 and y^2, are added.
- The coefficients of x^2 and y^2 are the same.

Examples of equations of circles:

Equivalent equation

$x^2 + y^2 = 30$	$x^2 + y^2 = \left(\sqrt{30}\right)^2$
$4x^2 + 4y^2 = 49$	$x^2 + y^2 = \left(\dfrac{7}{2}\right)^2$
$y^2 - 23 = 2 - x^2$	$x^2 + y^2 = 5^2$

Ellipses:

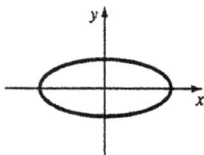

- Can be written in the form $\dfrac{x^2}{t^2} + \dfrac{y^2}{s^2} = 1$, t and $s > 0$.
- Both variables, x and y, are squared.
- The squared terms, x^2 and y^2, are added.
- The coefficients of x^2 and y^2 are not the same.

Examples of equations of ellipses:

Equivalent equation

$\dfrac{x^2}{9} + \dfrac{y^2}{16} = 1$	$\dfrac{x^2}{3^2} + \dfrac{y^2}{4^2} = 1$
$25y^2 + 4x^2 = 100$	$\dfrac{x^2}{5^2} + \dfrac{y^2}{2^2} = 1$
$4y^2 = 16 - x^2$	$\dfrac{x^2}{4^2} + \dfrac{y^2}{2^2} = 1$

131

Hyperbolas:

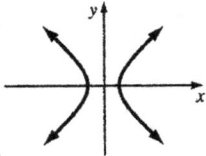

- Can be written in the form $\dfrac{x^2}{r^2} - \dfrac{y^2}{w^2} = 1$ or $\dfrac{y^2}{w^2} - \dfrac{x^2}{r^2} = 1$, r and $w > 0$.

- Both variables, x and y, are squared.

- The squared terms, x^2 and y^2, are not added.

Examples of equations of hyperbolas:

Equivalent equation

$\dfrac{x^2}{4} - \dfrac{y^2}{25} = 1$	$\dfrac{x^2}{2^2} - \dfrac{y^2}{5^2} = 1$
$4y^2 - 9x^2 = 36$	$\dfrac{y^2}{3^2} - \dfrac{x^2}{2^2} = 1$
$9x^2 - 225 = 25y^2$	$\dfrac{x^2}{5^2} - \dfrac{y^2}{3^2} = 1$

Parabolas:

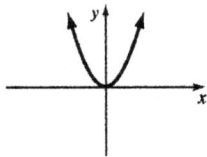

- Can be written in the form $y = ax^2 + bx + c$ or $x = ay^2 + by + c$.

- Only one of the variables, x and y, is squared.

Examples of equations of parabolas:

$y = 3x^2 + 13x - 10$

$x = y^2 - 5y + 4$

$3y = x^2$ $\left(\text{Equivalent equation: } y = \dfrac{1}{3}x^2\right)$

$y - 10x = x^2 + 24$ $\left(\text{Equivalent equation: } y = x^2 + 10x + 24\right)$

Examples Classify the equation as an equation of a circle, an ellipse, a hyperbola, or a parabola.

1. $6x^2 + 6y^2 = 216$
    - An equivalent equation is $x^2 + y^2 = 36$, or $x^2 + y^2 = 6^2$.
    - Both variables are squared, so this cannot be a parabola.
    - The squared terms are added, so this cannot be a hyperbola.
    - The coefficients of x^2 and y^2 are the same.
    - This is an equation of a circle.

2. $16x^2 + 25y^2 = 400$

  - An equivalent equation is $\dfrac{x^2}{25} + \dfrac{y^2}{16} = 1,$ or $\dfrac{x^2}{5^2} + \dfrac{y^2}{4^2} = 1.$
  - Both variables are squared, so this cannot be a parabola.
  - The squared terms are added, so this cannot be a hyperbola.
  - The coefficients of x^2 and y^2 are not the same.
  - This is an equation of an ellipse.

3. $49y^2 - 196 = 4x^2$

  - An equivalent equation is $\dfrac{y^2}{4} - \dfrac{x^2}{49} = 1,$ or $\dfrac{y^2}{2^2} - \dfrac{x^2}{7^2} = 1.$
  - Both variables are squared, so this cannot be a parabola.
  - The squared terms are not added, so this cannot be a circle or an ellipse.
  - This is an equation of a hyperbola.

4. $y - 7x = x^2 - 18$

  - An equivalent equation is $y = x^2 + 7x - 18.$
  - Only one variable is squared, so this cannot be a circle, an ellipse, or a hyperbola.
  - This is an equation of a parabola.

Exercises Classify the equation as an equation of a circle, an ellipse, a hyperbola, or a parabola.

1. $y = 3x^2$

2. $\dfrac{x^2}{100} + \dfrac{y^2}{25} = 1$

3. $\dfrac{x^2}{81} - \dfrac{y^2}{4} = 1$

4. $36y^2 - 9x^2 = 324$

5. $y^2 = 16 - x^2$

6. $x = y^2 + 6y - 27$

7. $y = x^2 - 2x$

8. $9x^2 + 4x^2 = 36$

9. $\dfrac{x^2}{16} - \dfrac{y^2}{25} = 1$

10. $5x^2 + 5y^2 = 20$

11. $y - 16x = x^2 + 55$

12. $4x^2 - 100 = 25y^2$

13. $25y^2 = 10 - x^2$

14. $\dfrac{1}{2}x^2 + \dfrac{1}{2}y^2 = 1$

15. $\dfrac{x^3}{3} + \dfrac{y^2}{5} = 1$

16. $x = y^2 + 4$

Interactive Preview 15: THE ELLIPSE

In this Preview, we will consider only ellipses with the center at the origin.

Example 1

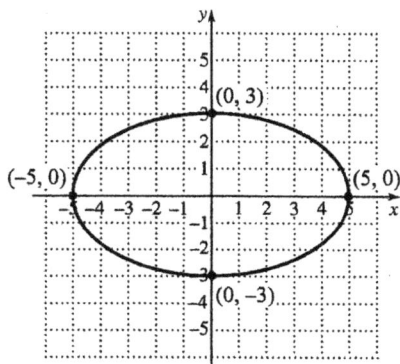

Equation: $\dfrac{x^2}{5^2} + \dfrac{y^2}{3^2} = 1$, or $\dfrac{x^2}{25} + \dfrac{y^2}{9} = 1$

$\underset{\rule{0pt}{0pt}}{\lfloor 5 > 3 \rfloor}$ and 5^2 is the denominator of $\dfrac{x^2}{5^2}$;

thus, the *major axis* is on the *x*-axis.

Center: $(0, 0)$

Endpoints of the **major axis**: $(-5, 0)$ and $(5, 0)$

These points are the **vertices** of the ellipse. They are also the *x-intercepts* of the ellipse. The major axis is *horizontal*.

Endpoints of the **minor axis**: $(0, -3)$ and $(0, 3)$

These points are the *y-intercepts* of the ellipse. The minor axis is *vertical*.

Example 2

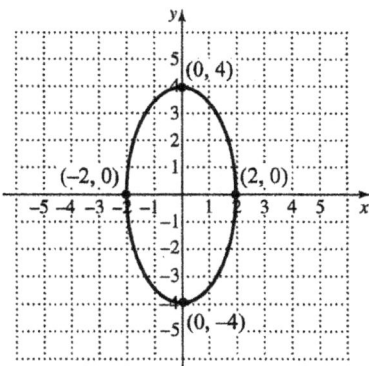

Equation: $\dfrac{x^2}{2^2} + \dfrac{y^2}{4^2} = 1$, or $\dfrac{x^2}{4} + \dfrac{y^2}{16} = 1$

$\underset{\rule{0pt}{0pt}}{\lfloor 4 > 2 \rfloor}$ and 4^2 is the denominator of $\dfrac{y^2}{4^2}$;

thus, the *major axis* is on the *y*-axis.

Center: $(0, 0)$

Endpoints of the **major axis**: $(0, -4)$ and $(0, 4)$

These points are the **vertices** of the ellipse. They are also the *y-intercepts* of the ellipse. The major axis is *vertical*.

Endpoints of the **minor axis:** $(-2, 0)$ and $(2, 0)$

These points are the *x-intercepts* of the ellipse. The minor axis is *horizontal*.

Graphs of Ellipses with Center at the Origin

- The longer axis is the **major axis**. The shorter axis is the **minor axis**.
- The endpoints of the major axis are the **vertices** of the ellipse.
- If the major axis is *horizontal*, the endpoints of the axis are the *x*-intercepts. Then the endpoints of the minor axis are the *y*-intercepts.
- If the major axis is *vertical*, the endpoints of the axis are the *y*-intercepts. Then the endpoints of the minor axis are the *x*-intercepts.

Exercises Fill in the coordinates of the *x*- and *y*-intercepts of the graph of the ellipse. Then fill in the blanks to complete the description of the graph.

1.

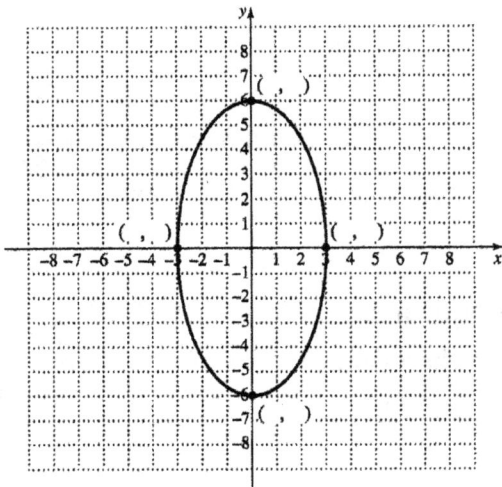

Center: (,)

The major axis is _____.
horizontal / vertical

The minor axis is _____.
horizontal / vertical

Vertices: (,) and (,)

Endpoints of major axis: (,) and (,)

Endpoints of minor axis: (,) and (,)

x-intercepts: (,) and (,)

y-intercepts: (,) and (,)

Equation: $\dfrac{x^2}{\square^2}+\dfrac{y^2}{\square^2}=1$, or $\dfrac{x^2}{\square}+\dfrac{y^2}{\square}=1$

2.

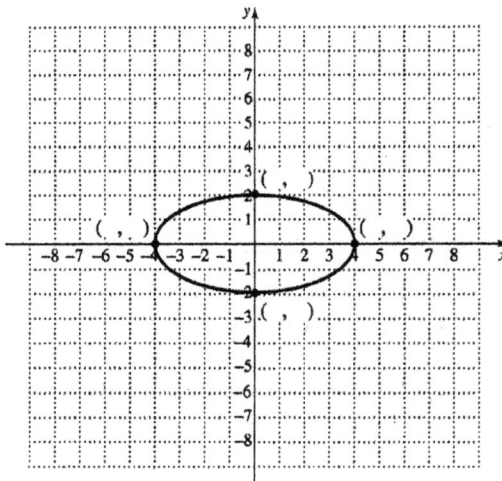

Center: (,)

The major axis is _____.
horizontal / vertical

The minor axis is _____.
horizontal / vertical

Vertices: (,) and (,)

Endpoints of major axis: (,) and (,)

Endpoints of minor axis: (,) and (,)

x-intercepts: (,) and (,)

y-intercepts: (,) and (,)

Equation: $\dfrac{x^2}{\square^2}+\dfrac{y^2}{\square^2}=1$, or $\dfrac{x^2}{\square}+\dfrac{y^2}{\square}=1$

Interactive Preview 16: THE HYPERBOLA

In this Preview, we will consider only hyperbolas with the center at the origin.

Example 1

Equation: $\dfrac{x^2}{4} - \dfrac{y^2}{25} = 1$, or $\dfrac{x^2}{2^2} - \dfrac{y^2}{5^2} = 1$

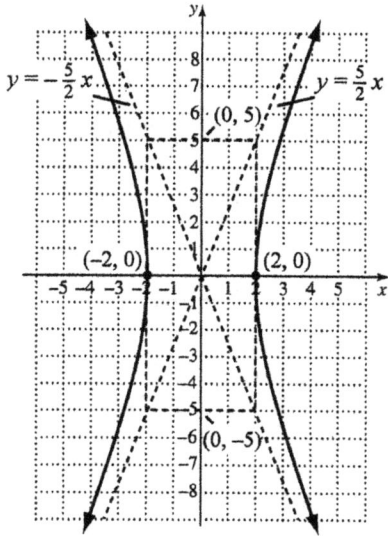

The x^2-term is the first term, thus
- the **transverse axis** is horizontal and is on the x-axis,
- the endpoints of the transverse axis are the **x-intercepts** $(-2, 0)$ and $(2, 0)$,
- the **vertices** of the hyperbola are the endpoints of the transverse axis, $(-2, 0)$ and $(2, 0)$, and
- the graph opens left and right.

The y^2-term is the second term, thus
- the **conjugate axis** is vertical and is on the y-axis, and
- the endpoints of the conjugate axis are $(0, -5)$ and $(0, 5)$.

The lines $y = \dfrac{5}{2}x$ and $y = -\dfrac{5}{2}x$ are the asymptotes of the hyperbola. As $|x|$ gets larger, the graph of the hyperbola gets closer and closer to the asymptotes.

Example 2

Equation: $\dfrac{y^2}{9} - \dfrac{x^2}{16} = 1$, or $\dfrac{y^2}{3^2} - \dfrac{x^2}{4^2} = 1$

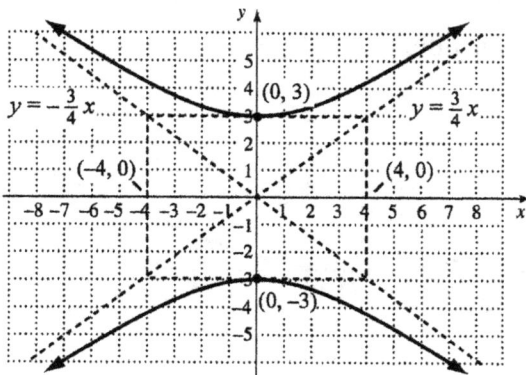

The y^2-term is the first term, thus
- the **transverse axis** is vertical and is on the y-axis,
- the endpoints of the transverse axis are the **y-intercepts** $(0, -3)$ and $(0, 3)$,
- the **vertices** of the hyperbola are the endpoints of the transverse axis, $(0, -3)$ and $(0, 3)$, and
- the graph opens up and down.

The x^2-term is the second term, thus
- the **conjugate axis** is horizontal and is on the x-axis, and
- the endpoints of the conjugate axis are $(-4, 0)$ and $(4, 0)$.

The lines $y = \dfrac{3}{4}x$ and $y = -\dfrac{3}{4}x$ are the asymptotes of the hyperbola. As $|y|$ gets larger, the graph of the hyperbola gets closer and closer to the asymptotes.

To draw a graph of a hyperbola, it helps to first sketch the asymptotes. The asymptotes of the hyperbola are the extended diagonals of the rectangle formed by sides passing through the endpoints of the transverse axis and the conjugate axis.

Example 3 Graph: $\dfrac{y^2}{16} - \dfrac{x^2}{49} = 1$, or $\dfrac{y^2}{4^2} - \dfrac{x^2}{7^2} = 1$.

The y^2-term is the first term. The transverse axis is vertical. The endpoints of the transverse axis are $(0, -4)$ and $(0, 4)$. The endpoints of the conjugate axis are $(-7, 0)$ and $(7, 0)$. We sketch a rectangle whose sides pass through these four points. (See Figure 1 below.)

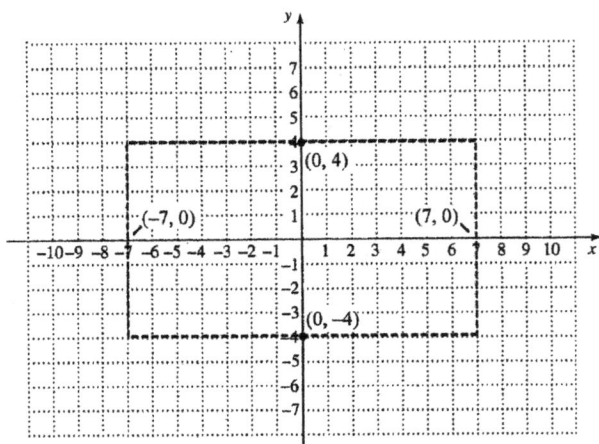

Figure 1

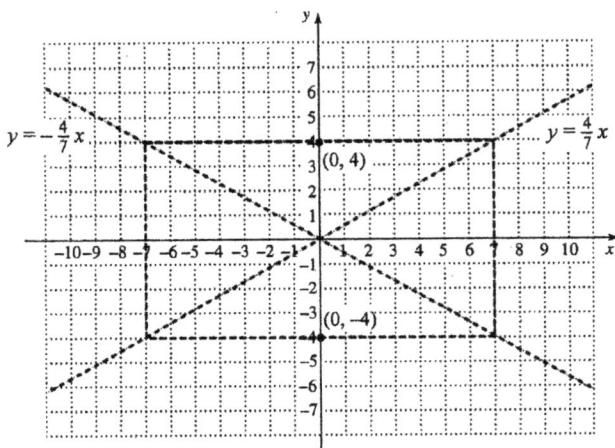

Figure 2

Next, we sketch the asymptotes which are the extended diagonals of the rectangle. (See Figure 2 above.) The equations of the asymptotes are $y = \dfrac{4}{7}x$ and $y = -\dfrac{4}{7}x$. The vertices $(0, 4)$ and $(0, -4)$ are on the y-axis. The graph opens up and down. Finally, we draw the branches of the hyperbola outward from the vertices toward the asymptotes. It can be helpful to plot four more points.

If $y = 5$ or $y - 5$, then

$$\frac{25}{16} - \frac{x^2}{49} = 1.$$

We solve for x:

$$-\frac{x^2}{49} = -\frac{9}{16}$$

$$x^2 = \frac{9 \cdot 49}{16}$$

$$x = \pm 5.25$$

Four points on the graph are $(5.25, 5)$, $(5.25, -5)$, $(-5.25, 5)$, and $(-5.25, -5)$.

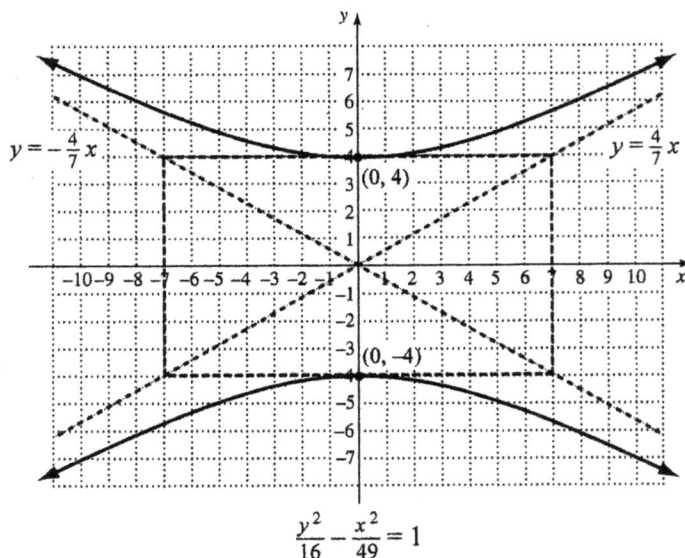

$$\frac{y^2}{16} - \frac{x^2}{49} = 1$$

Exercises

1. Graph: $\dfrac{x^2}{36} - \dfrac{y^2}{9} = 1$, or $\dfrac{x^2}{6^2} - \dfrac{y^2}{3^2} = 1$. Begin by completing the statements below.

    - The ___-term is the first term, thus the transverse axis is on the ___-axis.

    - The endpoints of the transverse axis are _____ and _____.

    - The endpoints of the conjugate axis are _____ and _____.

    - The vertices are _____ and _____.

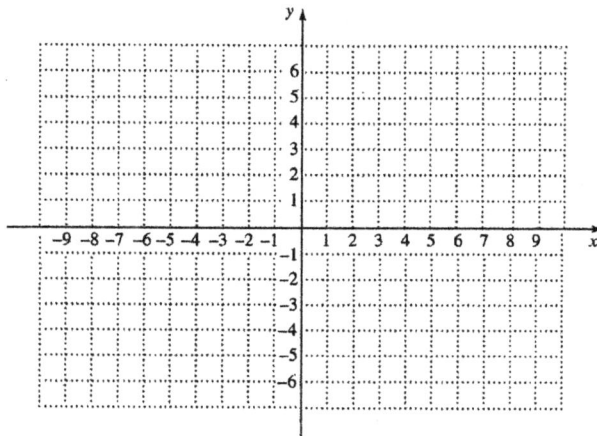

2. Graph: $\dfrac{y^2}{4} - \dfrac{x^2}{25} = 1$, or $\dfrac{y^2}{2^2} - \dfrac{x^2}{5^2} = 1$. Begin by completing the statements below.

    - The ___-term is the first term, thus the transverse axis is on the ___-axis.

    - The endpoints of the transverse axis are _____ and _____.

    - The endpoints of the conjugate axis are _____ and _____.

    - The vertices are _____ and _____.

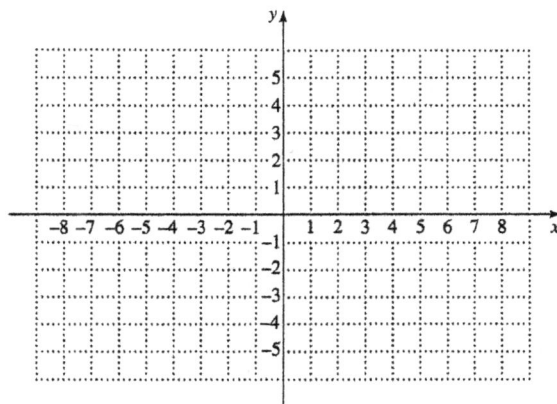

In Exercises 3-8, match each equation with its graph from choices A-F.

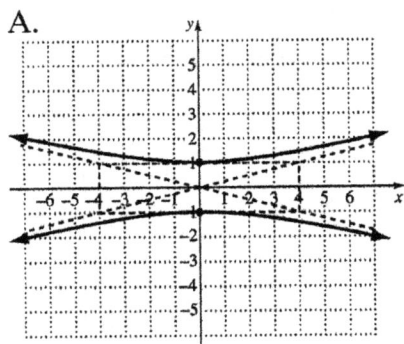

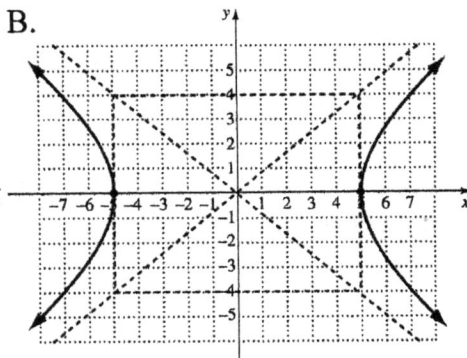

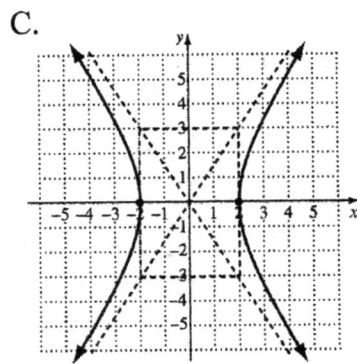

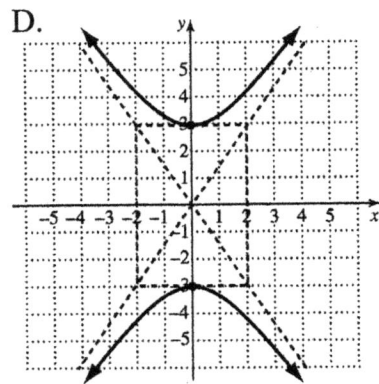

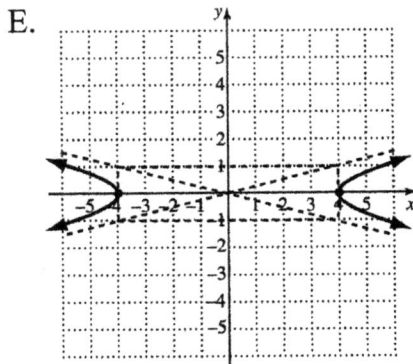

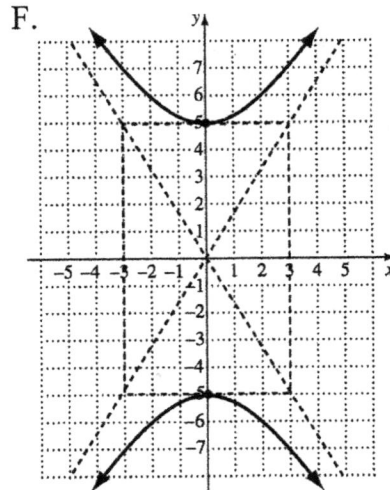

A.

B.

C.

D.

E.

F.

3. $\dfrac{x^2}{16} - \dfrac{y^2}{1} = 1$

4. $\dfrac{y^2}{9} - \dfrac{x^2}{4} = 1$

5. $\dfrac{y^2}{1} - \dfrac{x^2}{16} = 1$

6. $\dfrac{x^2}{25} - \dfrac{y^2}{16} = 1$

7. $\dfrac{y^2}{25} - \dfrac{x^2}{9} = 1$

8. $\dfrac{x^2}{4} - \dfrac{y^2}{9} = 1$

Student Activities

Correlation Guide

The **Student Activities** in this Notebook accompany *College Algebra*, 5th edition, by Beecher/Penna/Bittinger. The following table contains a correlation between the activity and the chapter in the text.

Student Activity #	Chapter in the Text
1	Just-In-Time 4
2	Just-In-Time 10
3	Just-In-Time 13
4	Just-In-Time 13
5	1
6	1
7	1
8	2
9	2
10	3
11	3
12	3
13	4
14	5
15	6
16	6
17	6
18	7
19	8

Activity 1: American Football

Focus: Order on the number line, addition with and without the number line, absolute value

American football is played on a field that is 100 yards long, not counting the end zones, and 53 1/3 yards wide. The yardage on a football field is marked as shown in the diagram. Each team's objective is to move the ball, by running with it or passing it, into the other team's end zone.

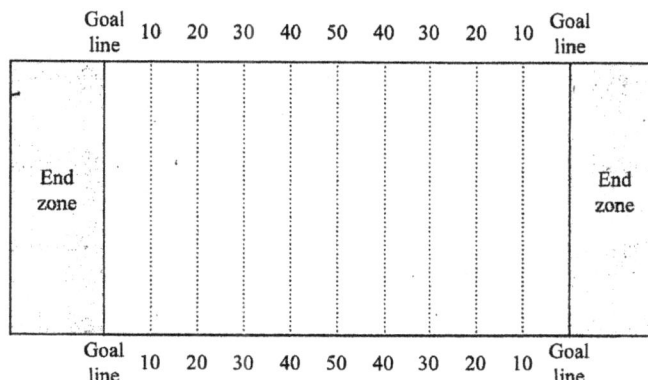

1. Draw a number line to model the length of the football field. Place 0 at the center of the field.

The football team with possession of the ball has 4 tries, called downs, to gain 10 yards. If it does not gain 10 yards after 4 downs, the other team wins possession of the ball.

On one possession, starting at the center of the field (location 0), a team heading in the positive direction on the number line made 4 plays in the following order:

      4-yard loss   6-yard gain   3-yard loss   12-yard gain

2. Use a number line to show where the ball was at the end of each play.

3. Where was the ball after the end of the 4 plays?

4. Did the team reach the goal of gaining 10 yards?

5. Write the final position of the ball as a sum.

Suppose the order of the plays was different:

      4-yard loss   6-yard gain   12-yard gain   3-yard loss

6. Write the final position of the ball as a sum.

7. How did the final position of the ball compare to the earlier result?

8. As soon as a team gains 10 yards, it earns 4 more downs to gain 10 more yards. Does this make any practical difference in the result of either of the previous sets of plays?

9. The distance the ball was moved on each play, not considering the direction of the move, is the **absolute value** of each number. For the game of football, is the direction of each play important, or is just the absolute value of each play important?

When a football team loses possession of the ball without scoring, it is called a turnover. There are two types of turnovers: turnovers due to fumbles or interceptions, and turnovers due to not advancing the ball 10 yards on 4 downs. Teams are ranked according to the following statistics:

- *Turnovers gained*: the number of times the team gains possession of the ball due to fumbles made by the other team plus the number of interceptions it makes

- *Turnovers lost*: the number of times the teams loses possession of the ball due to fumbles plus the number of interceptions made by the other team

- *Turnover margin*: the difference between turnovers gained and turnovers lost

10. If g = turnovers gained, l = turnovers lost, and m = turnover margin, write an equation that can be used to calculate turnover margin.

11. Is a positive turnover margin good or bad? Which turnover margin is better: -8 or 1?

12. Turnover margins for several NFL football teams during a recent season are listed. Draw a number line and graph the turnover margins, ranking the teams from best to worst.

Dallas Cowboys	6
Oakland Raiders	-15
Green Bay Packers	14
Pittsburgh Steelers	0
New York Giants	-2

Activity 2: Finding the Magic Number

Focus: Evaluating polynomials in several variables

Materials: A coin for each person

Note: This activity is designed for 3-member groups.

Can you determine the requirements for your baseball team to clinch first place?

A team's *magic number* is the combined number of wins by that team and losses by the second-place team that guarantee the leading team a first-place finish. For example, if the Cubs' magic number is 3 over the Reds, any combination of Cubs wins and Reds losses that totals 3 will guarantee a first-place finish for the Cubs. A team's magic number is computed using the polynomial

$$G - P - L + 1,$$

where G is the length of the season, in games, P is the number of games that the leading team has played, and L is the total number of games that the second-place team has lost minus the total number of games that the leading team has lost.

1. The standings below are from a fictitious league. Each group should calculate the Jaguars' magic number with respect to the Catamounts as well as the Jaguars' magic number with respect to the Wildcats. (Assume that the schedule is 162 games long.)

	W	L
Jaguars	92	64
Catamounts	90	66
Wildcats	89	66

2. Each group member should play the role of one of the teams, using coin tosses to simulate the remaining games. If a group member correctly predicts the side (heads or tails) that comes up, the coin toss represents a win for that team. Should the other side appear, the toss represents a loss. Assume that these games are against other (unlisted) teams in the league. Each group member should perform three coin tosses and then update the standings.

3. Recalculate the two magic numbers found in Question 1, using the updated standings from Question 2.

4. Slowly – one coin toss at a time – play out the remainder of the season. Record all wins and losses, update the standings, and recalculate the magic numbers each time all three group members have completed a round of coin tosses.

5. Examine the work in part (4) and explain why a magic number of 0 indicates that a team has been eliminated from contention.

Activity 3: Visualizing Factoring

Focus: Factoring trinomials

Materials: Graph paper and scissors

Factoring a polynomial like $x^2 + 5x + 6$ can be thought of as determining the length and the width of a rectangle that has area $x^2 + 5x + 6$.

1. a) To factor $x^2 + 11x + 10$ geometrically, first cut out shapes like those below to represent x^2, $11x$, and 10. This can be done by either tracing the figures below or by selecting a value for x, say 4, and using the squares on the graph paper to cut out the following:

   x^2: Using the value selected for x, cut out a square that is x units on each side.

   $11x$: Using the value selected for x, cut out a rectangle that is 1 unit wide and x units long. Repeat this to form 11 such strips.

   10: Cut out two rectangles with whole-number dimensions and an area of 10. One should be 2 units by 5 units and the other 1 unit by 10 units.

   b) Then attempt to use one of the two rectangles with area 10, along with all the other shapes, to piece together one large rectangle. Only one of the rectangles with area 10 will work.

   c) From the large rectangle formed in part (b), use the length and the width to determine the factorization of $x^2 + 11x + 10$. Where do the dimensions of the rectangle representing 10 appear in the factorization?

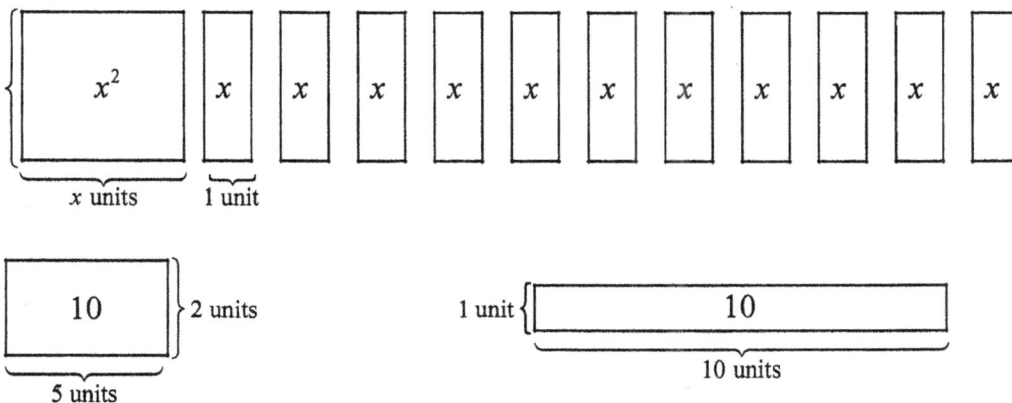

2. Repeat step (1) above, but this time use the other rectangle with area 10, and use only 7 of the 11 strips, along with the x^2-shape. Piece together the shapes to form one large rectangle. What factorization do the dimensions of this rectangle suggest?

3. Cut out rectangles with area 12 use the above approach to factor $x^2 + 8x + 12$. What dimensions should be used for the rectangle with area 12?

Activity 4: Matching Factorizations

Focus: Factoring polynomials

Materials: One set of cards, copied and cut out from the attached master, for each student or group of students.

Note: You will need a large surface such as a table for this activity.

The top portion of each of the cards in your set contains a polynomial, and the bottom portion contains a factorization. Shuffle the cards. Then choose one card and match the polynomial at the top of that card to the factorization at the bottom of a second card. Then match the polynomial at the top of the second card to the factorization at the bottom of a third card. Continue to match polynomials with their factorizations, forming a circle as you go. If your work is correct, the cards should form a complete circle when you are finished.

$x^2 - x - 6$	$x^2 + 5x + 6$	$2x^2 + 5x + 3$
$(x+2)(x+3)$	$(2x+3)(x+1)$	$(x-2)(x-3)$
$x^2 - 5x + 6$	$6x^2 + x - 1$	$x^2 + x - 6$
$(2x+1)(3x-1)$	$(x-2)(x+3)$	$(2x+3)(3x-2)$

$6x^2+5x-6$	$6x^2-5x-1$	$6x^2-5x-6$	$2x^2-5x+3$
$(6x+1)(x-1)$	$(2x-3)(3x+2)$	$(2x-3)(x-1)$	$(2x-1)(3x-1)$
$6x^2-5x+1$	$6x^2+13x+6$	$6x^2+5x-1$	$2x^2+x-3$
$(2x+3)(3x+2)$	$(6x-1)(x+1)$	$(2x+3)(x-1)$	$(2x-1)(3x+1)$
$6x^2-x-1$	$6x^2-7x+1$		
$(6x-1)(x-1)$	$(x+2)(x-3)$		

Activity 5: Pets in the United States

Focus: Percents, Formulas, Applications

Data: American Veterinary Medical Association

Although dogs and cats are the most popular pets in the United States, many American households contain other animals as well. The following table lists the number of households with at least one of seven types of pets and the total pet population.

	Number of Households (in thousands) in 2007	Number of Households (in thousands) in 2012	Total Pet Population (in thousands) in 2007	Total Pet Population (in thousands) in 2012
Dogs	43,021	43,346	72,114	69,926
Cats	37,460	36,117	81,721	74,059
Birds	4,453	3,671	11,199	8,300
Horses	2,087	1,780	7,295	4,856
Turtles	1,106	1,320	1,991	2,297
Snakes	390	555	586	1,150
Lizards	719	726	1,078	1,119

1. Familiarize yourself with the data. What information is given for each type of animal? The data is given "in thousands." What does that mean? For example, how many households had dogs in 2007?

2. For which pets did the number of households increase from 2007 to 2012? For which pets did the total pet population increase from 2007 to 2012? Why are these lists not the same?

3. If we consider only households containing a specific pet, then the formula $p = ah$ gives the pet population p given the average number of pets per household a and the number of households h that contain at least one pet. Solve the formula for a. For which pets listed in the table was the average number of pets per household greater than 2 in 2012?

4. Complete the following table. Use a + sign to indicate an increase and a − sign to indicate a decrease. Round percent change to the nearest tenth of a percent and number of households to the nearest one. (Recall that we find percent increase or decrease by computing $\dfrac{\text{amount of increase or decrease}}{\text{original amount}}$. For example, if a population increases from 200 to 225, the amount of increase is $225 - 200$, or 25, and the percent increase is $\dfrac{25}{200}$, or 0.125, or 12.5%.)

	Number of Households (in thousands) in 2007	Number of Households (in thousands) in 2012	Change in the Number of Households (in thousands)	Percent Change in the Number of Households
Dogs	43,021	43,346	+325	+ 0.8%
Cats	37,460	36,117	−1,343	−3.6%
Birds	4,453	3,671		
Horses	2,087	1,780		
Fish	9,036		−1,298	
Ferrets		334	−171	
Rabbits	1,870			−24.7%
Hamsters		877		+6.2%
Guinea Pigs		847	+219	
Gerbils	187			+25.1%
Turtles	1,106	1,320		
Snakes	390	555		
Lizards	719	726		

5. For which pet did the number of households increase the most from 2007 to 2012? For which pet was the percent increase in the number of households the greatest? Why are these not the same?

6. Estimate the number of households containing gerbils in 2017 in two ways. First, assume that the change in the number of households from 2012 to 2017 will be the same as the change from 2007 to 2012. Second, assume that the percent change in the number of households from 2012 to 2017 will be the same as the change from 2007 to 2012. Which method resulted in the larger estimation? Why?

7. The total population of fish kept as household pets in 2012 was ten times the population of snakes kept as household pets. How many fish were kept as household pets in 2012?

8. There were as many birds kept as household pets in 2012 as there were horses, hamsters, and turtles combined. What was the average number of hamsters per household in 2012?

9. Write a paragraph for a news source about pet ownership in the United States using some of the information from this activity. Include what you feel are the most interesting and important numbers you calculated and explain their relevance.

Activity 6: Linear Regression on a Graphing Calculator

Focus: Linear regression on a graphing calculator

Data: Nielsen

Due to the rise of streaming services such as Netflix, Amazon, and Hulu as well as a resurgence in video games, TV viewing among people ages 18 to 34 has declined at a fast pace in recent years. The following table gives the percentage of viewers in this age range watching television, including DVR playback, in an average minute in the period from September 24 to October 28 for the years 2014 through 2018.

Year, x	Percentage of Viewers, y
2014, 0	26.4
2015, 1	24.2
2016, 2	22.3
2017, 3	19.8
2018, 4	16.8

We will use a graphing calculator to fit a linear equation $y = mx + b$ to these data, where x is the number of years after 2014. (The keystrokes are for a TI-84 Plus calculator.) First we enter the data from the table on the STAT list screen. We will enter the x-values in L1 and the y-values in L2.

We begin by clearing any existing lists. To clear L1, move the cursor to the heading L1. Press CLEAR and then the down arrow. Do this for each of the other lists that contain entries.

Now position the cursor at the first blank space in L1 and enter the x-values. To enter 0, press 0 ENTER. Continue entering the x-values 1 through 4, each followed by ENTER. Press the right arrow to move to the top of the L2 column. Type the y-values in succession, each followed by ENTER.

To select linear regression from the STAT CALC menu and to store the equation as Y1 on the $Y =$ screen, press STAT right arrow 4 VARS right arrow 1 1 ENTER. The display will show the coefficients a and b of the regression line $y = ax + b$. (The values of r and r^2, which are indications of how well the equation fits the data, are also shown).

```
LinReg
 y=ax+b
 a=-2.36
 b=26.62
 r2=.9924447613
 r=-.9962152183
```

1. What does a represent in the regression equation? What does b represent?

2. Will the graph slant up from left to right or down? How do you know? Graph the equation in the window with $Xmin = 0$, $Xmax = 15$, $Yscl = 1$, $Ymin = 0$, $Ymax = 30$, and $Yscl = 5$ to confirm your answer.

3. Estimate the percentage of viewers ages 18 to 34 watching television in the given time period in 2019 and in 2020.

4. Predict when there will be no TV viewers age 18 to 34 in the given time period. Does this answer seem reasonable? Why or why not?

5. Data on the sales of frozen entrees, in billions of dollars, are given in the following table. Use a graphing calculator to fit a linear equation to the data. Let $x =$ the number of years after 2015.

Year, x	Sales of Frozen Entrees, y
2015, 0	$17.6
2016, 1	17.8
2017, 2	18.0
2018, 3	19.0

6. Estimate the sales of frozen entrees in 2020 and in 2021.

7. Research the reasons for this rise in sales in recent years.

8. Do you think this trend will continue? Does your research provide any insight into this?

Activity 7: Going Beyond High School

Focus: Circle, bar, and line graphs, percent, rate of change, equations of lines, graphing lines

Data: National Center for Education Statistics

Patterns of college attendance have changed significantly in the past few decades. The table below gives percentages of high school completers enrolling in college, for several demographics, for selected years between 1975 and 2013.

Year	Percent of Recent High School Completers Enrolled in a Two-Year College	Percent of Recent High School Completers Enrolled in a Four-Year College	Percent of Recent Male High School Completers Enrolled in College	Percent of Recent Female High School Completers Enrolled in College
1975	18.2	32.6	52.6	49.0
1980	19.4	29.9	46.7	51.8
1985	19.6	38.1	58.6	56.8
1990	20.1	40.0	58.0	62.2
1995	21.5	40.4	62.6	61.3
2000	21.4	41.9	59.9	66.2
2005	24.0	44.6	66.5	70.4
2010	26.7	41.4	62.8	74.0
2013	23.8	42.1	63.5	68.4

1. Examine the data in the table. What do the numbers represent? Why do you think that the column heads reference "high school completers" rather than "high school graduates"?

2. If there were 2,977,000 high school completers in 2013, how many enrolled in a college? How many enrolled in a two-year college? How many enrolled in a four-year college? How many did not enroll in college?

3. For 2013, make a graph to illustrate the percent of high school completers enrolling in two-year colleges, four-year colleges, and not in any college. What type of graph will best illustrate this data?

4. For 2013, make a graph to illustrate the number of high school completers enrolling in two-year colleges, four-year colleges, and not in any college, given that there were 2,977,000 high school completers in 2013. What type of graph will best illustrate this data?

5. Using one pair of axes, draw a graph to illustrate the percent of high school completers who enrolled in a two-year college and a graph to illustrate the percent of high school completers who enrolled in a four-year college for years from 1975 to 2013. What type of graph will best illustrate this data? What trends can you see from the graph?

6. Using data from 1975 and 2013, what is the rate of change for the percent of male high school completers who enrolled in college? Using data from the same years, what is the rate of change for the percent of female high school completers who enrolled in college? What do these rates mean? Do you think that these rates are good indicators of changes in college enrollment for high school completers between 1975 and 2013?

7. Use data from 1975 and 2013 to write an equation that could be used to model the percent of recent male high school completers who enrolled in college. Using data from the same years, write an equation that could be used to model the percent of recent female high school completers who enrolled in college. Using one pair of axes, graph both equations. Describe what these graphs tell you about these two categories of students.

8. Use regression and all the data from 1975 through 2013 to write an equation to model the percent of recent male high school completers who enrolled in college. Using regression and data from the same years, write an equation to model the percent of recent female high school completers who enrolled in college. Graph both equations in the same viewing window. Describe what these graphs tell you about these two categories of students. How do these equations differ from those found in Question 7?

9. Examine the data in the table shown below. For what categories in the table before Question 1 are percentages given here? How do these categories relate to the categories in that table?

Year	Percent of Recent Male High School Completers Enrolled in a Two-Year College	Percent of Recent Male High School Completers Enrolled in a Four-Year College	Percent of Recent Female High School Completers Enrolled in a Two-Year College	Percent of Recent Female High School Completers Enrolled in a Four-Year College
2010	28.5	34.3	24.6	49.3
2011	24.7	40.0	27.3	44.9
2012	26.9	34.4	30.7	40.6
2013	24.5	39.0	23.0	45.3

10. Using one pair of axes, draw a line graph for each of the four categories of students in the table in Question 9, using a different color for each category, if possible. Describe the patterns you see. What events in the United States in those years may have affected the college decisions of high school completers?

11. In 2016, a high school guidance counselor wants to know what her ninth grade students will be doing after graduation. Using data and equations from this activity, write a paragraph describing the overall picture of the types of colleges high school completers enroll in and what she can expect her students to do.

Activity 8: Data and Downloading

Focus: Direct variation, inverse variation

Although download speed and data plan limits continue to increase, so do the size of the files consumers download. The table below lists a typical file size for several types of files, along with the time it takes to download one of those files and the number of that type of file that could be downloaded within a 5 GB data limit. The units used are related as follows.

1 MB (megabyte) = 1000 KB (kilobyte)
1 GB (gigabyte) = 1000 MB = 1,000,000 KB

File Type	File Size	Time It Takes to Download a File at 10 Mbps (megabytes per second) Download Speed	Maximum Number of Downloads in One Month with 5 GB Monthly Plan Limit
Email without attachments	5 KB (0.005 MB)	0.0005 sec	1,000,000
Word document	100 KB (0.1 MB)	0.01 sec	50,000
High-resolution image	5 MB	0.5 sec	1000
2-minute YouTube video	10 MB	1 sec	500
3-hour movie	1000 MB	100 sec	5
1-hour HDTV show	2500 MB	250 sec	2

1. Explore the data. As the file size increases, does the time it takes to download that type of file increase or decrease? As the file size increases, does the number of downloads possible of that type of file increase or decrease?

2. Divide the time it takes, in seconds, to download each type of file by the file size, in MB. What is the common ratio? Considering the units of this ratio, what does the ratio represent?

3. Multiply each of the file sizes, in MB, by the corresponding maximum number of downloads possible of that type of file. What is the common product? What does it represent?

4. Graph the data. On one grid, place file size on the horizontal axis and download time on the vertical axis. On a second grid, place file size on the horizontal axis and number of downloads on the vertical axis.

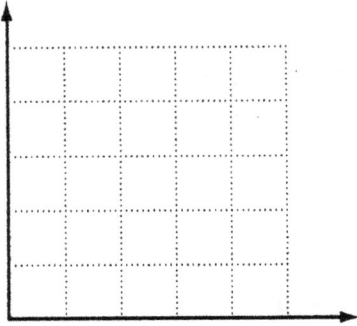

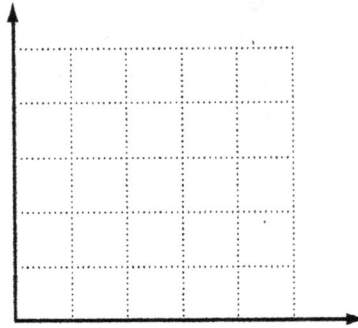

5. Which two quantities appear to vary directly? Which two quantities appear to vary inversely?

6. Find an equation of variation that describes how download time t varies with file size x. How is the constant in this equation related to the common ratio found in Question 2?

7. Find an equation of variation that describes how the number of downloads n varies with file size x. How is the constant in this equation related to the common product found in Question 3?

8. A 2-minute MP3 music file is about 2 MB. Use one of the equations from the table to determine how long it would take to download a 2-minute song. Use one of the equations above to determine how many 2-minute songs could be downloaded within a 5 GB data limit.

9. Write a paragraph comparing direct variation and inverse variation.

Activity 9: Mobile Data

Focus: Algebra of functions; percent increase

Data: Cisco Visual Networking Index: Forecast and Methodology, 2014–2019.

Consumer use of mobile devices to send and receive data and to access the internet is growing rapidly. The growth, however, does not follow the same pattern in all regions of the world. The table below lists projections for the average monthly mobile data and internet traffic, in petabytes.

(A petabyte is 1×10^{15} bytes.)

Region	2014	2015	2016	2017	2018	2019	Amount of Increase 2014 to 2019	Percent Increase 2014 to 2019
Asia Pacific, A	977	1622	2616	4114	6245	9459	8482	868%
North America, N	563	849	1287	1897	2704	3798	3235	575%
Central and Eastern Europe, C	242	464	832	1409	2231	3488	3246	
Middle East and Africa, M	199	383	690	1194	1927	3051		
Western Europe, W	341	504	760	1137	1653	2392		
Latin America, L	201	354	581	915	1380	2032		
Total, T	2523	4176						

1. Explore the data. The regions are listed in order of their projected average monthly mobile data usage in 2019. Is this order the same as it was in 2014? If not, how is the order different?

2. Complete the table, filling in the Total row and the Amount of Increase and Percent Increase columns. Which region saw the greatest amount of increase? Which saw the greatest percent increase?

3. In the table, a function name is given after each region name. We can refer to entries in the table using those function names. For example, $A(2014)$ represents the average monthly mobile data usage in the Asia Pacific region in 2014, so $A(2014) = 977$. What is the value of each of the following, and what quantity does it represent?

   (a) $W(2019)$
   (b) $(C + W)(2014)$
   (c) $(N - L)(2019)$
   (d) $(A + N + C + M + W + L)(2019)$

4. By stacking the regional monthly mobile data for each year, we can show regional and total data in the same graph. The data for 2014 to 2017 are graphed below. Complete the graph through 2019. If you have colored pencils available, shade the regions between each pair of lines. What does the height of each shaded region represent?

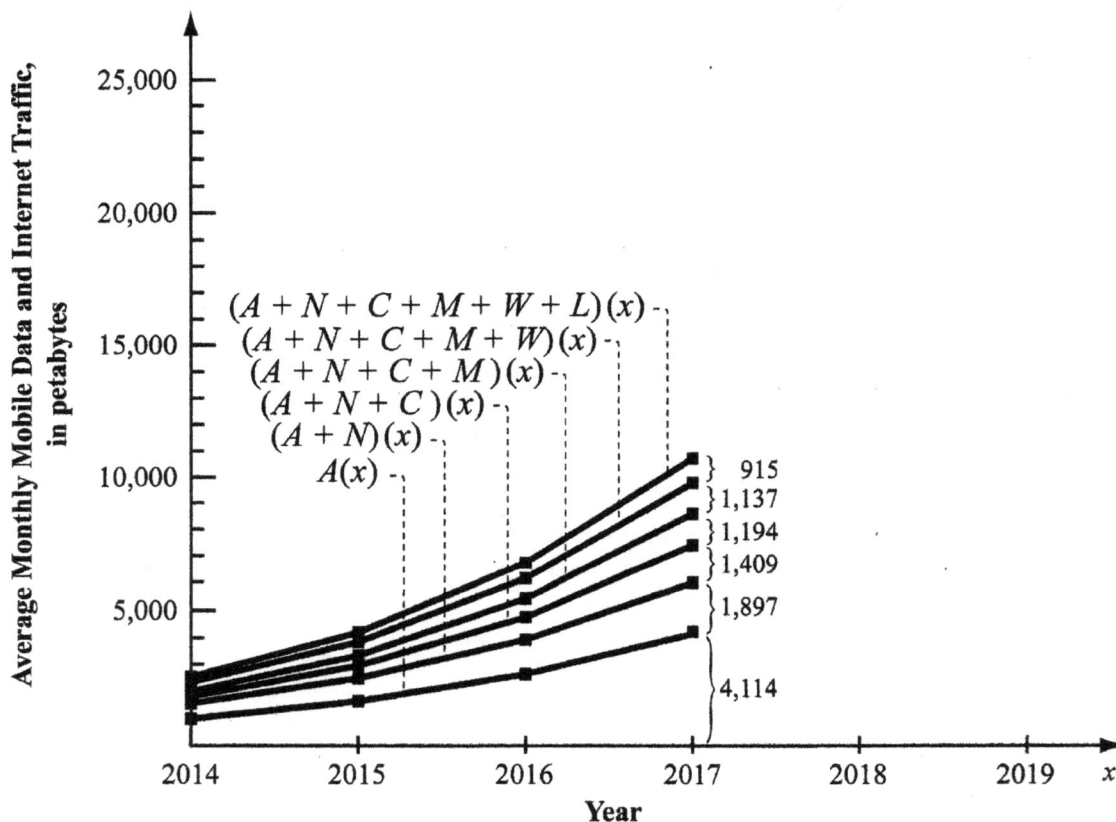

5. Why is mobile data usage predicted to grow at different rates across the regions? List two or three factors that might affect the growth of mobile data usage, and explain how they might differ across these geographic regions.

Activity 10: Firefighting Formulas

Topics: Evaluating radical expressions, graphing radical equations, simplifying
      radical expressions

Data for Formulas: Manchaca Fire Rescue (mvfd.org) and
                     Truckee Meadow Fire Protection District (tmfire.us)

In order to effectively fight fires, firefighters rely on accurate estimations of amount and
velocity of water flow. Accuracy and safety also depend on water pressure in the hose and
the nozzle reaction force. The following table lists water flow for selected nozzle pressures
for a $1\frac{3}{4}$-inch diameter smooth-bore nozzle tip.

Nozzle pressure, NP, in pounds per square inch	Water flow, Q, in gallons per minute	Nozzle reaction force, in pounds
20	410	
40	575	
60	705	
80	815	
100	910	
120	1000	
140	1075	

1. Explore the data. As nozzle pressure increases, does water flow increase or decrease?
   Does the relationship appear to be linear?

2. Graph the data.

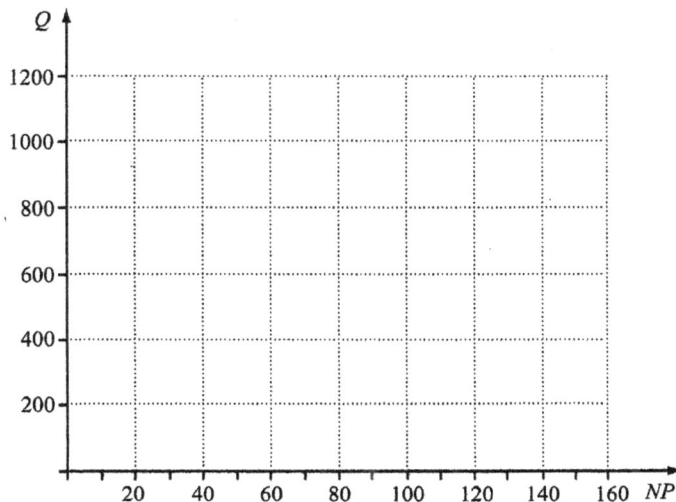

Firefighters use a number of formulas to calculate nozzle reaction, water flow, and volume. Several of these formulas are listed below.

$$Q = 29.72D^2\sqrt{NP}$$ Equation 1

$$NR = 1.57D^2 \times NP \text{ (for a solid stream)}$$ Equation 2

$$V = 12.1\sqrt{NP}$$ Equation 3

where

Q = water flow, in gallons per minute
V = water velocity, in feet per second
D = nozzle diameter, in inches
NR = nozzle reaction force, in pounds
NP = nozzle pressure, in pounds per square inch

3. Graph the equation $Q = 29.72D^2\sqrt{NP}$, where $D = 1.75$, on the same graph as the data given in the table. Does the equation appear to be a good fit?

4. Equation 2 gives nozzle reaction in terms of diameter and nozzle pressure. Use Equation 1 and Equation 2 to find a formula that gives nozzle reaction in terms of water flow and nozzle pressure. (*Hint:* Solve Equation 1 for D^2.)

5. Suppose, for a $1\frac{1}{2}$-inch nozzle tip, the nozzle pressure is 60 pounds per square inch. Find the water flow, the nozzle reaction force, and the water velocity.

6. Use Equation 2 to fill in the nozzle reaction force for a $1\frac{3}{4}$-inch nozzle tip in the third column of the table on the preceding page.

7. As nozzle pressure increases, both water flow and nozzle reaction force increase. An increase in water flow means more water can be used to extinguish a fire. Is an increase in nozzle reaction force helpful or harmful for firefighters? Research nozzle reaction force, and explain what it is and how it affects firefighting.

Activity 11: Music Downloads

Focus: Linear and quadratic models

Data: Recording Industry Association of America

Sales of physical recorded music, such as CDs, declined as digital sales increased. In time, digital music sales also began to decrease. The table below lists sales of albums and single tracks, in millions of units, for years between 2008 and 2014.

Year	Number of Digital Albums Sold, in millions	Number of Digital Single Tracks Sold, in millions
2008	428	1070
2009	379	1159
2010	326	1172
2011	331	1271
2012	316	1336
2013	289	1259
2014	257	1108

1. Explore the data. How many digital albums were sold in 2008? In 2014? How many digital single tracks were sold in 2008? In 2014?

2. Looking at only the data from 2008 and 2014, did the number of digital albums sold increase or decrease? Looking at only the data from 2008 and 2014, did the number of digital single tracks increase or decrease?

3. Let x represent the number of years after 2008, and graph the number of digital albums sold for all years from 2008 to 2014. How would you describe the trend of the data? What type of function would you use to model the data?

4. Let x represent the number of years after 2008, and graph the number of digital single tracks sold for all years from 2008 to 2014. How would you describe the trend of the data? What type of function would you use to model the data?

5. Choose two points from the digital album data and fit a linear function to the data. Graph this line with the data. How well does the function model the data? Is there a pair of points that you think would have resulted in a better model?

6. Choose three points from the digital single tracks data and fit a quadratic function to the data. Graph this parabola with the data. How well does the function model the data? Is there another set of three points that you think would have resulted in a better model?

7. Using all the points and a graphing calculator, fit a linear function to the digital album data. Graph this line with the data. How does the fit of this function compare with the function you found in Question 5?

8. Using all the points and a graphing calculator, fit a quadratic function to the digital single tracks data. Graph this line with the data. How does the fit of this function compare with the function you found in Question 6?

9. If the trends continue, will the number of single tracks sold ever equal the number of digital albums sold? If so, when? How might these trends influence the way artists record and market their music?

10. In what year were the sales of single tracks at a maximum? What do you think caused the sales of single tracks to decline after that year? In order to answer this question, you may need additional information. What other data would you like to collect, and why?

Activity 12: Let's Go to the Movies

Focus: Modeling with polynomial functions

Data: National Association of Theatre Owners

As the following table indicates, the number of annual movie admissions in the United States and Canada increased for several years after 1992, then began to decrease.

Year, x	Number of Movie Admissions in the United States and Canada, in billions
1992, 0	1.099
1994, 2	1.24
1996, 4	1.319
1998, 6	1.438
2000, 8	1.383
2002, 10	1.57
2004, 12	1.484
2006, 14	1.401
2008, 16	1.341
2010, 18	1.339
2012, 20	1.36

1. Examine the data. For what years did admissions increase? For what years did admissions decrease? What trends do you see in the data?

2. Graph the data on the grid below or on a graphing calculator. Choose a scale that shows the trends. Would you use a linear function or a quadratic function to model the data?

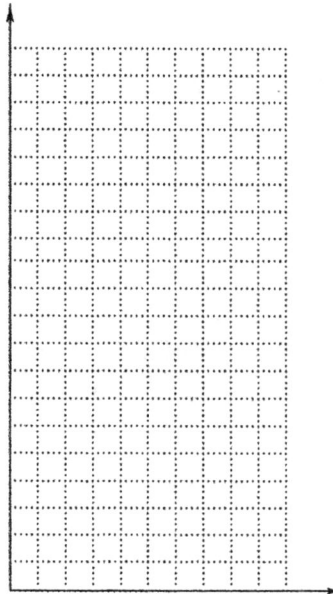

3. Fit a quadratic function to the data, either algebraically by choosing three points or using a graphing calculator and quadratic regression. Label this function E.

4. Now look at just the years from 1992 to 2002. If this were all the data you had, would you use a linear function or a quadratic function to model the data? Fit a function to this first part of the data set either algebraically or with regression using a graphing calculator. Label this function F.

5. Now look at just the years from 2002 to 2012. If this were all the data you had, would you use a linear function or a quadratic function to model the data? Fit a function to this second part of the data set either algebraically or with regression using a graphing calculator. Label this function S.

6. The number of movie admissions in 2014 was 1.27 billion. Would any of your three functions E, F, or S have predicted this number? Which one was the closest?

7. Research to find the number of movie admissions in the United States and Canada for the latest year possible. Would any of your three functions E, F, or S have predicted this number? Which one was the closest?

8. Write a paragraph describing the results of your analysis. Include your thoughts about what might have caused the overall increasing and decreasing trends in movie admissions that you observed.

Activity 13: Teaching About Zeros

Focus: Theorems about zeros of polynomial functions

Materials: Phone or computer for recording video

Group size: 2-4

One of the best ways to know that you understand a concept is to be able to explain it to someone else. Choose one of the following questions, or use the one assigned to you by your instructor, and use your textbook or other resources to thoroughly examine theorems that answer the question. Then create a video that answers the question, teaching the concepts in such a way that your classmates will also understand the question and the answer. Your video should be less than five minutes in length and should contain at least the following:

- Description and examples of the types of polynomial functions you are discussing
- Clear answers to the question using language that is both accurate and understandable
- At least one worked out example; more may be necessary if your result has several parts
- At least one exercise for viewers to try on their own

When all videos are complete, play yours for the class and check the answers to the exercises you requested your classmates to solve.

1. What do you know about complex number solutions of polynomial functions with real coefficients?

2. What do you know about irrational zeros of polynomial functions with rational coefficients?

3. What do you know about rational zeros of polynomial functions with integer coefficients?

4. What can you tell about the number of positive real zeros and the number of negative real zeros from the coefficients of a polynomial function?

Suggestions for videos:

- Use a phone to record yourself teaching at a blackboard or whiteboard.

- Use a smart pen to record your explanation as you write it on paper or on a tablet.

- Prepare a slide presentation on your computer and record audio as you work through the presentation.

- Prepare a presentation on your computer and make a screencast.

- Use educational software to make a video.

Activity 14: Earthquake Magnitude

Focus: Applications of logarithms

One way of describing the size of an earthquake is by estimating its *seismic moment* or the force required to produce the waves recorded on a seismograph. Rather than actually giving the seismic moment M_0 (measured in dyne·cm), scientists often instead list an earthquake's *moment magnitude* M_W, where

$$M_W = \frac{2}{3}\log(M_0) - 10.7.$$

The following table lists the moment magnitude for significant earthquakes in Chile in 2014 and 2015.

Date	Moment Magnitude	Seismic Moment
March 16, 2014	7.0	
April 1, 2014	8.2	
April 1, 2014	7.5	
April 1, 2014	7.0	
April 2, 2014	7.7	
August 23, 2014	6.4	
October 8, 2014	7.0	
March 18, 2015	6.3	
March 23, 2015	6.4	
September 16, 2015	8.3	

1. Explore the data. Which was the strongest earthquake listed? The weakest? Can you tell from the moment magnitude how much greater the strongest earthquake was than the weakest?

2. Solve the equation $M_W = \frac{2}{3}\log(M_0) - 10.7$ for M_0. Use this new equation to find the seismic moment for each earthquake in the table.

3. We can divide the seismic moments of two earthquakes to find out how many times stronger one earthquake was than another. How many times stronger was the September 16, 2015, earthquake than the March 18, 2015, earthquake?

4. The September 16, 2015, earthquake was 2 units of moment magnitude greater than the March 18, 2015, earthquake. Will the proportional moment magnitude of any two earthquakes that are 2 units apart be the same as the proportional seismic moment of these two earthquakes?

5. Suppose two earthquakes are one unit of moment magnitude apart. Use the equation $M_W = \frac{2}{3}\log(M_0) - 10.7$ to determine how many times stronger the stronger earthquake is than the weaker earthquake.

6. Calculate the seismic moment of earthquakes with moment magnitude 1, 2, 3, 4, 5, 6, 7, 8, and 9. Which would be easier to illustrate on a graph: all 9 moment magnitudes or all 9 seismic moments?

7. Why might scientists use moment magnitudes instead of seismic moments to describe the strength of an earthquake?

Activity 15: Fruit Juice Consumption

Focus: Systems of equations and graphing

Data: U. S. Department of Agriculture

The table below lists the per capita consumption of various fruit juices in the United States, in gallons per person, for several years.

Year	Orange Juice	Apple Juice and Cider	Grape Juice	Grapefruit Juice	Pineapple Juice
1985	4.81	1.55	0.23	0.61	0.34
1990	3.25	1.74	0.28	0.91	0.50
1995	4.73	1.59	0.45	0.59	0.38
2000	5.54	1.80	0.34	0.53	0.30
2005	4.76	1.87	0.51	0.23	0.26
2010	3.45	2.21	0.37	0.22	0.23
2013	4.01	1.78	0.43	0.20	0.21

1. Explore the data. For which juices does per capita consumption appear to be increasing? Decreasing? Neither?

2. Graph each set of data using a line graph. Which patterns appear to be linear?

3. Focus on grape juice, grapefruit juice, and pineapple juice. Fit a linear equation to each set of data, and graph the lines along with the data.

4. Do the slopes of the lines support your initial analysis of whether per capita consumption is increasing or decreasing?

5. Examine the points of intersection of the three lines. What does each point tell you about the consumption of these juices?

6. Now focus on orange juice and apple juice and cider. Fit a linear equation to each set of data, and graph the lines along with the data. What is the predicted per capita consumption of orange juice in 2015? What is the predicted per capita consumption of apple juice and cider in 2015? In what year will per capita orange juice consumption be equal to per capita apple juice and cider consumption?

7. Which of the lines that you fit for the five juices appear to be a good fit for the data?

The previous data looked at long-term trends of per capita consumption of several fruit juices. The table below lists per capita consumption for the same juices, in gallons per person, for every year from 2000 to 2010.

Year	Orange Juice	Apple Juice and Cider	Grape Juice	Grapefruit Juice	Pineapple Juice
2000	5.54	1.80	0.34	0.53	0.30
2001	5.15	1.79	0.33	0.54	0.31
2002	5.02	1.80	0.37	0.47	0.32
2003	4.09	1.95	0.40	0.40	0.34
2004	4.94	2.13	0.38	0.38	0.27
2005	4.76	1.87	0.51	0.23	0.26
2006	4.39	2.22	0.44	0.20	0.27
2007	4.13	2.28	0.56	0.29	0.22
2008	3.79	2.11	0.45	0.30	0.27
2009	3.92	2.09	0.38	0.26	0.27
2010	3.45	2.21	0.37	0.22	0.23

8. Explore the data. Do any of the trends seem different for this decade than they did for the years from 1985 to 2013?

9. Graph each set of data. Which patterns appear to be linear?

10. Now focus on orange juice and apple juice and cider. Fit a linear equation to each set of data, and graph the lines along with the data. What is the predicted per capita consumption of orange juice in 2015? What is the predicted per capita consumption of apple juice and cider in 2015? In what year will per capita consumption of orange juice be equal to per capita consumption of apple juice and cider?

11. Were your predictions the same for Questions 6 and 10? If they were different, which prediction do you think is the most reliable? Find the actual values for 2015 for both orange juice and apple juice and cider consumption. How close were your estimates? What factors, other than historic trends, might a producer of juice consider when trying to predict demand?

Activity 16: Waterfalls

Focus: Linear inequalities

1. The Western New York Waterfall
    Society has compiled
    classifications of waterfalls by
    their shape. Write and graph an
    inequality that describes each of
    the types of waterfalls.

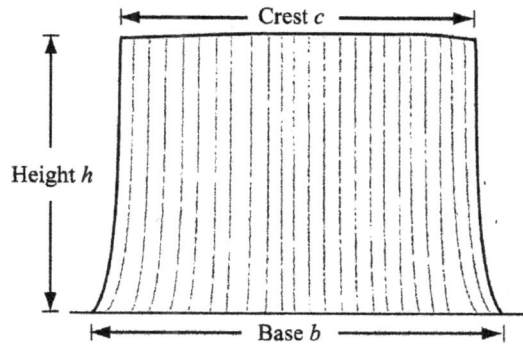

Height h

Crest c

Base b

Fan: The crest width c is less than
or equal to one-third the base
width b.

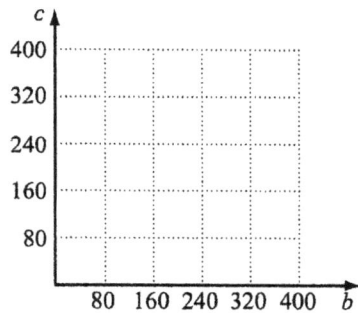

c
400
320
240
160
80

80 160 240 320 400 b

Funnel: The crest c is greater than
three times the base width b.

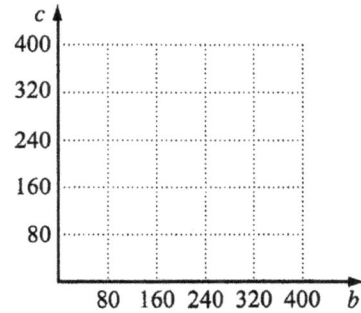

c
400
320
240
160
80

80 160 240 320 400 b

Curtain: The crest width c is
greater than one and one-
half times the height h.

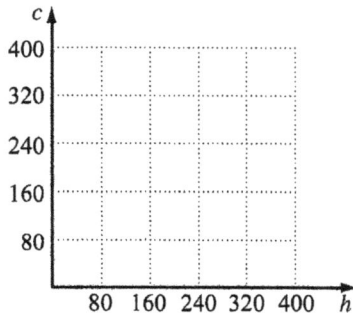

c
400
320
240
160
80

80 160 240 320 400 h

Ribbon: The crest width c is less
than or equal to one-half
the height h.

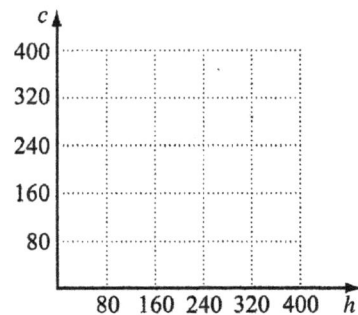

c
400
320
240
160
80

80 160 240 320 400 h

2. Each of the falls described below can be plotted as a point on a graph. Locate each of the waterfalls described below on the graph of each inequality above. What is the most appropriate classification for each waterfall?

   (a) Angeline Falls
   King County, Washington
   Height: 400 ft
   Crest width: 100 ft
   Base width: 100 ft

   (b) Brotherhood Falls
   Skagit County, Washington
   Height: 66 ft
   Crest width: 5 ft
   Base width: 40 ft

   (c) Lower Lewis River Falls
   Skamania County, Washington
   Height: 43 ft
   Crest width: 200 ft
   Base width: 200 ft

Activity 17: Construction

Focus: Linear inequalities; systems of linear inequalities

Guardrails for worker safety at construction sites are required to be 42 inches high. In some cases, it is impossible to build guardrails to this height. According to the Oregon Occupational Health and Safety Administration, for these sites it is acceptable to build a shorter barrier that meets the following two requirements:
- The barrier must be a minimum of 30 inches high.
- The sum of the height and depth of the barrier must be at least 48 inches.

1. Which of the following barriers meets these requirements?

(a)

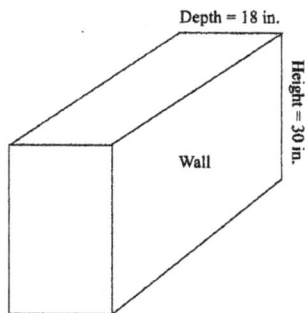

Depth = 18 in.
Height = 30 in.
Wall

(b)

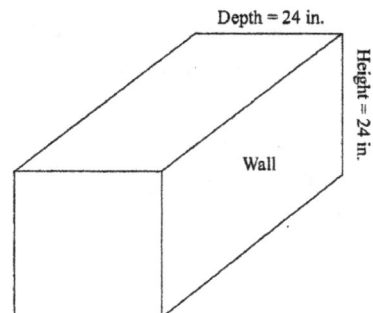

Depth = 24 in.
Height = 24 in.
Wall

(c)

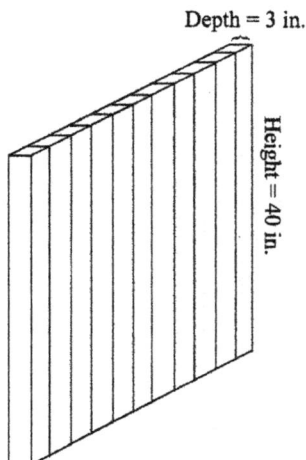

Depth = 3 in.
Height = 40 in.

(d)

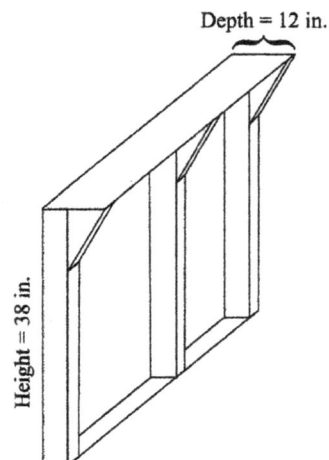

Depth = 12 in.
Height = 38 in.

2. Write a linear inequality that models the requirement that the barrier must be a minimum of 30 inches high. Let h represent the height of the barrier. Graph the inequality on the grid below.

3. Write a linear inequality that models the requirement that the sum of the height and depth of the barrier must be at least 48 inches. Let h represent the height of the barrier and d represent the depth of the barrier. Graph the inequality on the same grid, using a different shading.

You have now graphed a system of inequalities. The overlap in the shading shows the solution set of the system.

4. Plot each of the barriers described above as a point on the grid. How does this graph support your answer to question 1?

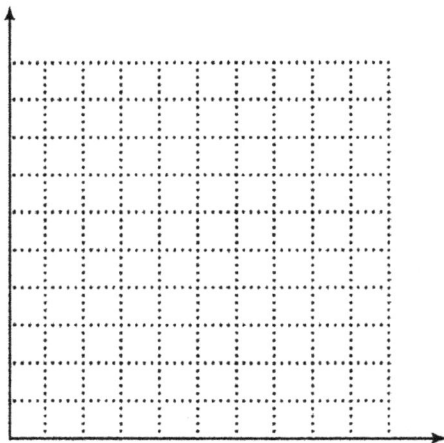

Activity 18: A Cosmic Path

Focus: Ellipses

On May 4, 2007, Comet 17P/Holmes was at the point closest to the sun in its orbit. Comet 17P is traveling in an elliptical orbit with the sun as one focus, and one orbit takes about 6.88 years. One astronomical unit (AU) is 93,000,000 mi. One group member should do the following calculations in AU and the other in millions of miles.

Data: Harvard-Smithsonian Center for Astrophysics

1. At its *perihelion*, a comet with an elliptical orbit is at the point in its orbit closest to the sun. At its *aphelion*, the comet is at the point farthest from the sun. The perihelion distance for Comet 17P is 2.053218 AU, and the aphelion distance is 5.183610 AU. Use these distances to find a. (See the following diagram.)

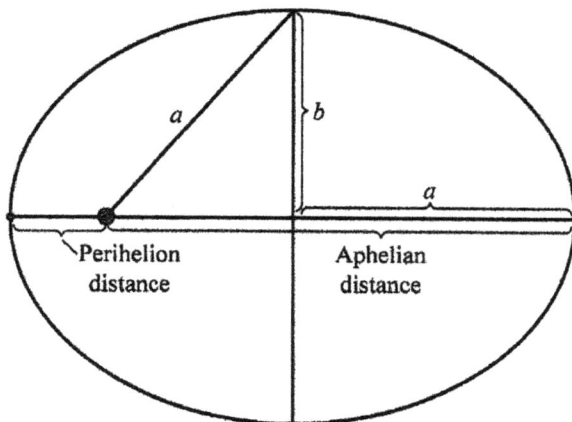

2. Using the figure above, express b^2 as a function of a. Then find b using the value found for a in part (1).

3. One formula for approximating the perimeter of an ellipse is

$$P = \pi\left(3a + 3b - \sqrt{(3a+b)(a+3b)}\right),$$

developed by the Indian mathematician S. Ramanujan in 1914. How far does comet 17P travel in one orbit?

4. What is the speed of the comet? Find the answer in AU per year and in miles per hour.

5. Which calculations – AU or mi – were easier to use? Why?

Activity 19: Bargaining for a Used Car

Focus: Geometric series

Time: 30 minutes

Group size: 2

Materials: Calculators

1. One group member (the "seller") is asking $5500 for a car. The second member (the "buyer") offers $1500. The seller splits the difference ($5500 - $1500 = $4000, and $4000 ÷ 2 = $2000) and lowers the price to $3500 ($5500 - $2000 = $3500). The buyer splits the difference again ($3500 - $1500 = $2000, and $2000 ÷ 2 = $1000) and gives a counter offer of $2500 ($1500 + $1000 = $2500). Continue in this manner until you are able to agree on the car's selling price to the nearest penny.

2. Check several guesses to find what the buyer's initial offer should be in order to achieve a purchase price of $2500 or less.

3. The seller's price in the bargaining above can be modeled recursively by the sequence

$$a_1 = 5500, \qquad a_n = a_{n-1} - \frac{d}{2^{2n-3}},$$

where d is the difference between the initial price and the first offer. Use this recursively defined sequence to solve parts (1) and (2) above.

4. The first four terms in the sequence in part (3) can be written as

$$a_1, \quad a_1 - \frac{d}{2}, \quad a_1 - \frac{d}{2} - \frac{d}{8}, \quad a_1 - \frac{d}{2} - \frac{d}{8} - \frac{d}{32}.$$

Use the formula for the limit of an infinite geometric series to find a simple algebraic formula for the eventual sale price, P, when the bargaining process from above is followed. Verify the formula by using it to solve parts (1) and (2) above.

This activity is based on the article "Bargaining Theory, or Zeno's Used Cars," by James C. Kirby, *The College Mathematics Journal*, 27(4), September, 1996.

Answers: Review Worksheets

Integrated Review 1:   GRAPHING LINEAR EQUATIONS

Check Your Understanding

1.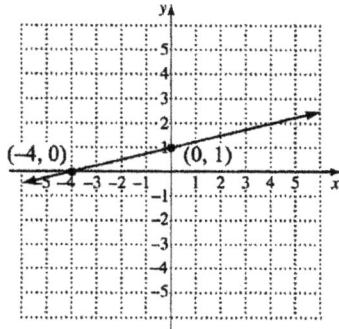

2.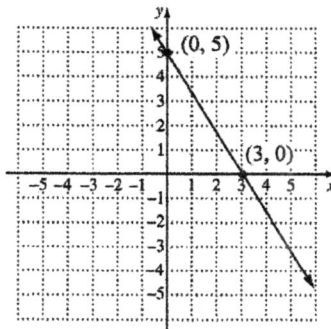

3.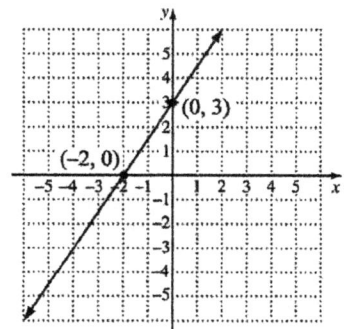

4. False   **5.** True   **6.** True   **7.** C   **8.** B   **9.** D   **10.** A

Exercises

1.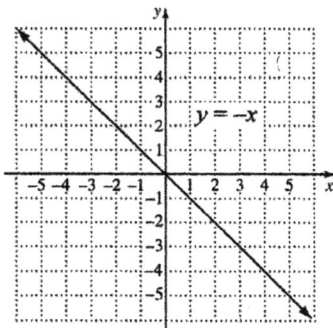

2.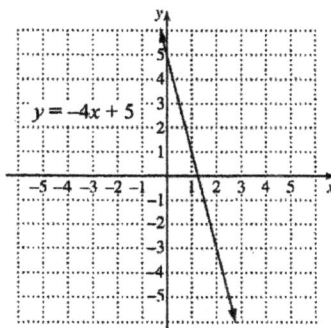

3.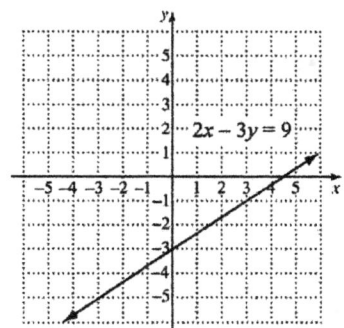

4.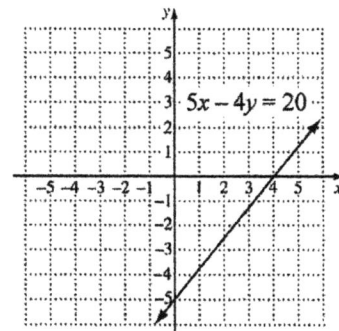

5.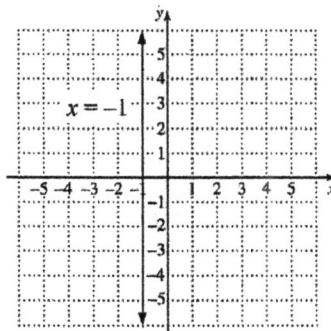

6.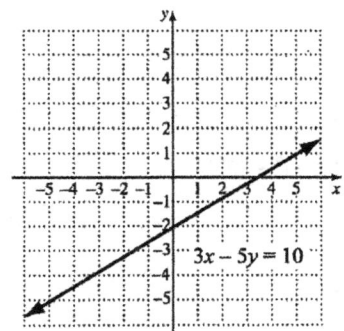

7.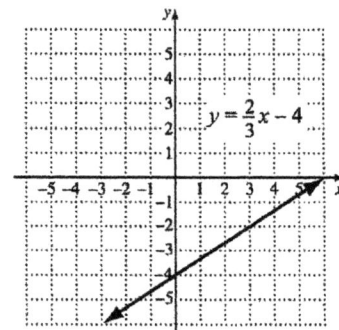

8.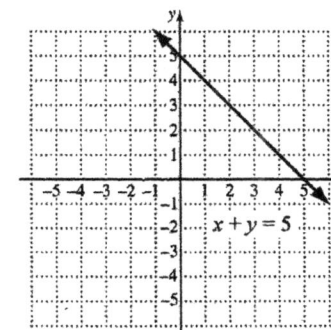

9.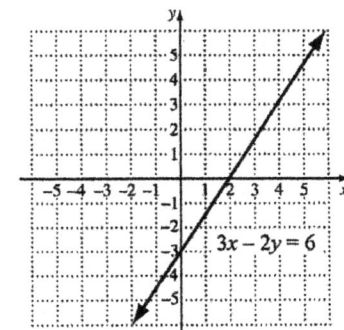

10.

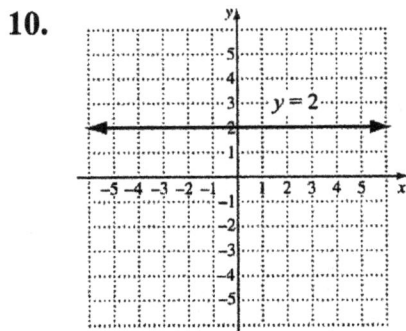

$y = 2$

11.

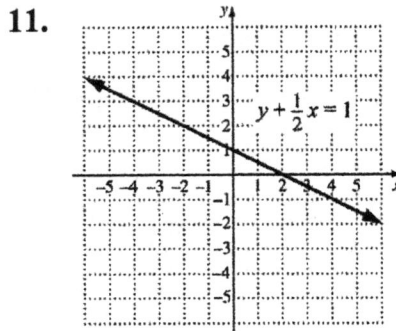

$y + \frac{1}{2}x = 1$

12.

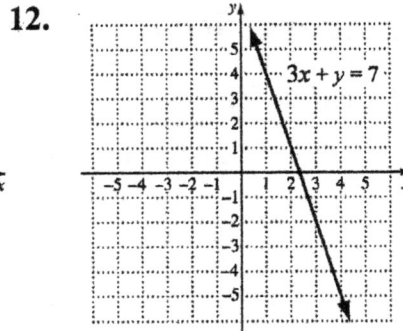

$3x + y = 7$

Integrated Review 2:   OPERATIONS ON THE REAL NUMBERS

Check Your Understanding

1. C **2.** E **3.** H **4.** B **5.** G

6. D **7.** A **8.** F

Exercises

1. 14 **2.** −5 **3.** 240 **4.** −2 **5.** $-\frac{9}{4}$

6. $-\frac{2}{25}$ **7.** −4.84 **8.** −8 **9.** 8 **10.** −34

11. −70 **12.** $-\frac{10}{9}$ **13.** $\frac{36}{11}$ **14.** −5.4 **15.** $\frac{85}{24}$

16. $-\frac{1}{16}$ **17.** −27 **18.** $\frac{2}{3}$ **19.** −2.04 **20.** 28

Integrated Review 3:   ORDER OF OPERATIONS

Check Your Understanding

1. A; in Solution B, Step (1) should have been dividing 18 by 6 instead of adding $6+3$.

2. B; in Solution A, Step (4) should have been dividing 80 by 5 instead of multiplying $5(8)$.

Exercises

1. Evaluate 2^2.
   Multiply $3 \cdot 5$.
   Divide $8 \div 4$.
   Subtract $6 - 15$.
   Add $-9 + 2$.

2. Subtract $5 - 3$.
   Evaluate 2^2.
   Divide $32 \div 4$.
   Multiply $8 \cdot 4$.
   Multiply $2 \cdot 2$.
   Subtract $32 - 4$.

3. −13 **4.** 60 **5.** 4 **6.** 4 **7.** 16

8. 187 **9.** −11 **10.** 70 **11.** 22 **12.** 6

13. 27 **14.** 292 **15.** 222 **16.** 12 **17.** −105

18. 139 **19.** 36 **20.** 216

Integrated Review 4:   FACTORING TRINOMIALS: $ax^2 + bx + c$

Exercises

1. $(x+9)(x-4)$

2. $(y-1)(y-9)$

3. $(y-5)(y+2)$

4. $(s+6)(s-15)$

5. $(x+8)(x-1)$

6. $(b+6)(b+2)$

7. $(x-8)(x+7)$

8. $(a+6)(a-9)$

9. $(y-7)(y+2)$

10. $(w-11)(w+1)$

11. $(q+9)(q+10)$

12. $(x-4)(x-10)$

13. $(x+20)(x+100)$

14. $(z-5)(z-5)$

15. $(3x-2)(2x+7)$

16. $(2w+3)(2w-5)$

17. $(5x+2)(6x-1)$

18. $(3y-4)(8y+1)$

19. $(2x+5)(10x-3)$

20. $(8c+3)(5c+2)$

21. $x(x-3)(x-2)$

22. $b(b+4)(b-8)$

23. $2y^2(y+15)(y-2)$

24. $5q(q+9)(q+4)$

Integrated Review 5:   FACTORING TRINOMIAL SQUARES AND DIFFERENCES OF SQUARES

Check Your Understanding

1. Difference of squares

2. Neither

3. Trinomial square

4. Difference of squares

5. Neither

6. Trinomial square

Exercises

1. $(8x+1)(8x-1)$

2. $(x-6)^2$

3. $(9y+4)(9y-4)$

4. $(w-2)^2$

5. $(7x+3)^2$

6. $(5a+6)(5a-6)$

7. $(7+z)(7-z)$

8. $(x+15)(x-15)$

9. $(y+8)^2$

10. $(2x-5)^2$

11. $5(x+3)(x-3)$

12. $3(x+4)^2$

Integrated Review 6:   THE PRINCIPLE OF ZERO PRODUCTS

Check Your Understanding

1. False   2. True   3. False

Exercises

1. $x^2 - 7x = -10$

$x^2 - 7x + 10 = 0$

$(x-5)(x-2) = 0$

$x - 5 = 0 \quad or \quad x - 2 = 0$

$x = 5 \quad or \quad x = 2$

The solutions are 5 and 2.

2. $15z^2 = -3z$

$15z^2 + 3z = 0$

$3z(5z+1) = 0$

$3z = 0 \quad or \quad 5z + 1 = 0$

$z = 0 \quad or \quad z = -\dfrac{1}{5}$

The solutions are 0 and $-\dfrac{1}{5}$.

3. $7, -9$

4. $0, \dfrac{1}{2}$

5. $-5, -4$

6. -10

7. $-4, 8$

8. $-2, -\dfrac{3}{4}$

9. $\dfrac{2}{3}, -\dfrac{5}{7}$

10. $0, -2, 3$

Integrated Review 7:   INTERVAL NOTATION

Check Your Understanding

1. D   **2.** C   **3.** C   **4.** A   **5.** B   **6.** A

Exercises

1. $(-\infty, 6)$

2. $[-4.1, 5.6)$

3. $(-8, 21]$

4. $[3, \infty)$

5. $(0, \infty)$

6. $(-\infty, \infty)$

7. $[0, 5]$

8. $(-6, 3)$

9. $\left(-\infty, \dfrac{2}{5}\right)$

10. $[17, \infty)$

11. $\left[-\dfrac{1}{2}, \dfrac{3}{2}\right)$

12. $(-3, -2]$

13.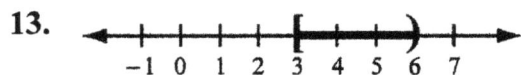

14.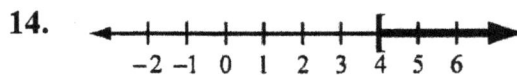

15.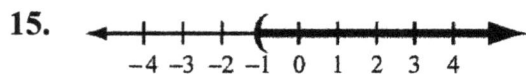

16.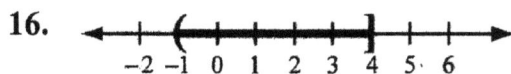

17.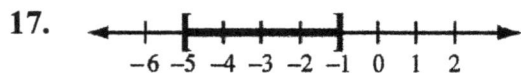

18.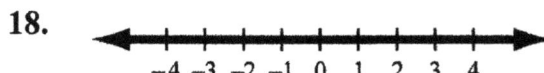

19.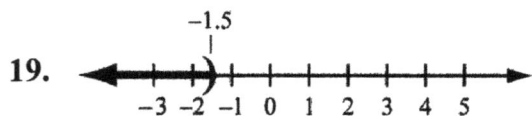

20.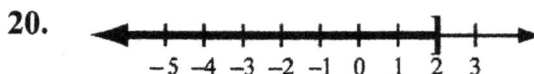

21.

```
◄──┼──┣━━━┿━━━┿━━━┿━━━┿━━━┿━━━┿━)──┼──►
   -14 -12 -10  -8  -6  -4  -2   0   2
```

22.

```
◄──┼──┼──┼──┼──(━━━┿━━━┿━)──┼──┼──┼──►
  -30 -20 -10   0   10  20  30  40  50
```

23.

```
◄──┼──┼──(━━━┿━━━┿━━━┿━━━┿━━━┿━━━┿━━━►
  -20 -15 -10  -5   0   5   10  15  20
```

24.

```
        -3                    5
        |                     |
◄──┼──┼──(━━━┿━━━┿━━━┿━━━┿━━┫──┼──┼──►
  -8  -6  -4  -2   0   2   4   6   8
```

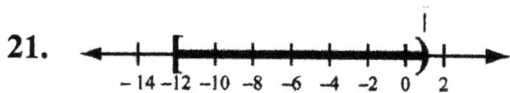

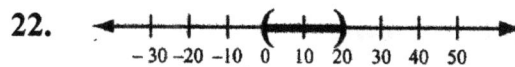

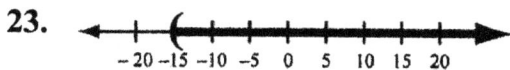

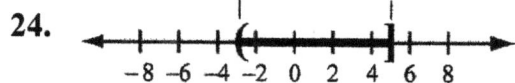

Integrated Review 8:   SOLVING LINEAR EQUATIONS

Exercises

1. 3 **2.** 1 **3.** −32 **4.** 147 **5.** −2.5

6. −6 **7.** −12 **8.** 22 **9.** 13 **10.** $\dfrac{27}{2}$

11. −5 **12.** 5 **13.** 2 **14.** 21 **15.** $\dfrac{18}{5}$

16. $\dfrac{10}{3}$ **17.** $-\dfrac{1}{10}$, or -0.1 **18.** 6 **19.** $\dfrac{3}{10}$

20. 3120 **21.** −4 **22.** $-\dfrac{28}{27}$

Integrated Review 9:   SOLVING LINEAR INEQUALITIES

Exercises

1. $\{y \mid y \geq -3\}$, or $[-3, \infty)$

2. $\left\{x \mid x > \dfrac{31}{20}\right\}$, or $\left(\dfrac{31}{20}, \infty\right)$

3. $\{c \mid c < -6\}$, or $(-\infty, -6)$

4. $\{q \mid q \geq 6\}$, or $[6, \infty)$

5. $\left\{x \mid x < -\dfrac{2}{3}\right\}$, or $\left(-\infty, -\dfrac{2}{3}\right)$

6. $\{z \mid z < 12\}$, or $(-\infty, 12)$

7. $\{x \mid x \geq 1\}$, or $[1, \infty)$

8. $\{x \mid x \leq 5\}$, or $(-\infty, 5]$

9. $\{t \mid t \leq 4\}$, or $(-\infty, 4]$

10. $\{x \mid x > -4\}$, or $(-4, \infty)$

Integrated Review 10:   THE PYTHAGOREAN THEOREM

Exercises

1. $b = 8$ **2.** $c = 17$ **3.** $a = 5$

4. $c = \sqrt{32} \approx 5.657$ **5.** $b = \sqrt{120} \approx 10.954$ **6.** $a = \sqrt{175} \approx 13.229$

7. $c = \sqrt{34} \approx 5.831$ **8.** $b = \sqrt{132} \approx 11.489$ **9.** $a = \sqrt{75} \approx 8.660$

10. $c = \sqrt{98} \approx 9.899$

Integrated Review 11: SIMPLIFYING RATIONAL EXPRESSIONS

Exercises

1. $\dfrac{4}{5}$

2. $3w$

3. $\dfrac{x+4}{9x+1}$

4. $\dfrac{x+10}{x-9}$

4. $\dfrac{1}{9y^6}$

6. $\dfrac{4z-5}{z+15}$

7. $\dfrac{y-7}{y+4}$

8. $\dfrac{2}{x}$

7. $\dfrac{a+2}{a-8}$

10. $\dfrac{8}{y}$

11. -1

12. $\dfrac{3x-1}{2x+1}$

Integrated Review 12: ADDING AND SUBTRACTING POLYNOMIALS

Check Your Understanding

1. f   2. l   3. i   4. c

Exercises

1. $-2t^2 - 6t - 3$

2. $4x + 5$

3. $9x^2 - 3x - 16$

4. $20y^2 + 4y + 6$

5. $10b^3 - b^2 - 11b - 3$

6. $3x^4 - 5x^3 - 11x^2 + x - 4$

7. $16x^2 + 9x - 1$

8. $2a^4 + 7a^3 - 9a^2 - 4a + 7$

9. $-10x^4 - 4x^3 + 2x - 8$

10. $w^5 - 5w^3 + 2w^2 + 4$

Integrated Review 13: SIMPLIFYING COMPLEX RATIONAL EXPRESSIONS

Exercises

1. $\dfrac{y}{14}$

2. $\dfrac{3}{2}$

3. $\dfrac{7}{20}$

4. $\dfrac{88}{15}$

5. $\dfrac{1-9x}{1+2x}$

6. $\dfrac{3z-5}{4z+3}$

7. $\dfrac{p^2-5}{p^2+5}$

8. $\dfrac{1+7a^2}{1+a^2}$

9. $\dfrac{2x-1}{x}$

10. $\dfrac{y}{5y+1}$

11. $\dfrac{ab}{b-a}$

12. $\dfrac{-1}{z(z+h)}$, or $-\dfrac{1}{z(z+h)}$

13. $\dfrac{-2}{x(x+h)}$, or $-\dfrac{2}{x(x+h)}$

Integrated Review 14: MULTIPLYING BINOMIALS

Exercises

1. $y^2 - 5y + 6$

2. $16b^2 - 38b - 5$

3. $w^2 + 3w - 10$

4. $z^2 - 8z - 9$

5. $20x^2 - 9x - 18$

6. $3q^2 + 65q + 100$

7. $x^2 + 9x - 52$ **8.** $30b^2 + 19b - 28$ **9.** $25w^2 + 35w - 8$

10. $a^2 + 17a + 66$ **11.** $x^2 + 12x + 36$ **12.** $4x^2 - 4x + 1$

13. $9a^2 + 12a + 4$ **14.** $t^2 + 20t + 100$ **15.** $z^2 - 24z + 144$

16. $16w^2 + 40w + 25$ **17.** $b^2 - 6b + 9$ **18.** $36y^2 - 36y + 9$

19. $y^2 - 100$ **20.** $9c^2 - 16$ **21.** $t^2 - 36$

22. $w^2 - 1$ **23.** $4z^2 - 49$ **24.** $100y^2 - 25$

25. $x^2 + 7x - 30$ **26.** $16x^2 - 81$ **27.** $21x^2 - 19x + 4$

28. $y^2 + 25y + 100$ **29.** $s^2 - 9$ **30.** $a^2 - 16a + 64$

31. $25y^2 + 40y + 16$ **32.** $6a^2 + 64a - 22$

Integrated Review 15: SIMPLIFYING RADICAL EXPRESSIONS
Exercises

1. $2\sqrt{3}$ **2.** $5\sqrt{3}$ **3.** $4\sqrt{5}$ **4.** $6\sqrt{2}$

5. $11\sqrt{3}$ **6.** $15\sqrt{2}$ **7.** $8\sqrt{5}$ **8.** $10\sqrt{6}$

9. $5\sqrt{x}$ **10.** y^{11} **11.** $6x$ **12.** $a + 1$

13. $2t^2\sqrt{21}$ **14.** $x - 5$ **15.** $w^4\sqrt{w}$ **16.** $4x\sqrt{2}$

17. $x - 12$ **18.** $(y+1)\sqrt{3}$ **19.** $10b^3\sqrt{14}$ **20.** $9x^3\sqrt{x}$

Integrated Review 16: MULTIPLYING RADICAL EXPRESSIONS
Exercises

1. $3\sqrt{6}$ **2.** $10\sqrt{3}$ **3.** $15\sqrt{6}$ **4.** $6\sqrt{7x}$

5. $6\sqrt{6a}$ **6.** 13 **7.** $23\sqrt{y}$ **8.** $8w\sqrt{15}$

9. $2t$ **10.** $z + 9$ **11.** $\sqrt{28x + 7}$ **12.** $50\sqrt{10}$

13. $9x\sqrt{2}$ **14.** $x^6\sqrt{x}$ **15.** y^8

Integrated Review 17: COMPLETING THE SQUARE
Check Your Understanding

1. $x^2 + 8x + 16 = 3 + 16$ **2.** $y^2 - 6y + 9 = -1 + 9$

3. $t^2 - 5t + \dfrac{25}{4} = -4 + \dfrac{25}{4}$ **4.** $x^2 + x + \dfrac{1}{4} = 2 + \dfrac{1}{4}$

Exercises

1. $1 \pm \sqrt{2}$

2. $-7 \pm \sqrt{23}$

3. $-\dfrac{5}{2} \pm \dfrac{\sqrt{37}}{2}$

4. $\dfrac{1}{2} \pm \dfrac{\sqrt{33}}{2}$

5. $4 \pm \sqrt{13}$

6. $-3 \pm \sqrt{5}$

7. $-\dfrac{11}{2} \pm \dfrac{\sqrt{129}}{2}$

8. $-\dfrac{3}{4} \pm \dfrac{\sqrt{89}}{4}$

Integrated Review 18: FIND THE LCM OF ALGEBRAIC EXPRESSIONS

Exercises

1. 270

2. 144

3. $9x^3$

4. $4(x-6)(x+6)$

5. $a(a-8)(a+8)$

6. $(y+3)(y-10)(y+10)$

7. $(x+2)(x-2)(x+3)$

8. $10t^3(t-4)$

9. $z^7(z+1)(z-1)^2$

10. $(a-7)(a+3)(a+1)$

11. $210(x-1)(x+1)$

12. $9y^3(8y-9)$

13. $18q^6(q-6)(q+1)$

14. $70c^2(c+7)(c-5)$

Integrated Review 19: RAISING RADICALS TO POWERS

Exercises

1. 6

2. 2

3. $x+7$

4. $y-8$

5. $3z+1$

6. $x+4\sqrt{x}+4$

7. $16-8\sqrt{3x}+3x$

8. $y-3+2\sqrt{y-4}$

9. $14-6\sqrt{2t+5}+2t$

Integrated Review 20: INTRODUCTION TO POLYNOMIALS

Exercises

1. No

2. Yes

3. Yes

4. Yes

5. No

6. No

7.

Term	$3y^4$	$-6y^3$	$8y^2$	$-\dfrac{2}{3}y$	5
Degree of Term	4	3	2	1	0
Degree of Polynomial	4				
Leading Term	$3y^4$				
Leading Coefficient	3				
Constant Term	5				

8.

Term	x	$-3x^4$	$4x^3$	-13	$7x^2$	$-6x^5$
Degree of Term	1	4	3	0	2	5
Degree of Polynomial	5					
Leading Term	$-6x^5$					
Leading Coefficient	-6					
Constant Term	-13					

9. Binomial **10.** Trinomial **11.** Monomial

12. Trinomial **13.** Monomial **14.** Binomial

15. $2y^3 - 8y^2 + y;\ y - 8y^2 + 2y^3$ **16.** $-a^4 + a^3 + 3;\ 3 + a^3 - a^4$

17. $-t^3 + 3t^2 - 5t + 3;\ 3 - 5t + 3t^2 - t^3$ **18.** $8x^5 - 4x^3 + 12x + 5;\ 5 + 12x - 4x^3 + 8x^5$

19. $-9b^3 - 5b^2 + 4b + 23;\ 23 + 4b - 5b^2 - 9b^3$

20. $5q^7 + \dfrac{3}{5}q - \dfrac{1}{3};\ -\dfrac{1}{3} + \dfrac{3}{5}q + 5q^7$

Integrated Review 21:   FACTORING BY GROUPING

Checking Your Understanding

1. Yes   **2.** No   **3.** No   **4.** Yes

Exercises

1.
$$y^3 + 4y^2 + 9y + 36$$
$$= \left(y^3 + 4y^2\right) + \left(9y + 36\right)$$
$$= y^2(y + 4) + 9(y + 4)$$
$$= (y + 4)\left(y^2 + 9\right)$$

2.
$$3s^3 - 15s^2 - 7s + 35$$
$$= 3s^3 - 15s^2 + \left(-7s + 35\right)$$
$$= 3s^2(s - 5) - 7(s - 5)$$
$$= (s - 5)\left(3s^2 - 7\right)$$

3. $(c + 6)\left(c^2 - 5\right)$ **4.** $(x - 7)\left(x^2 + 4\right)$ **5.** Cannot be factored by grouping.

6. $2(y - 1)\left(y^2 + 9\right)$ **7.** $(x - 5)\left(x^2 + 1\right)$ **8.** $(2y + 3)\left(y^2 - 3\right)$

9. $(y - z)(x + w)$ **10.** $\left(x^2 + 3\right)\left(2x^2 - 5\right)$

Integrated Review 22:   INTEGERS AS EXPONENTS

Exercises

1. $\dfrac{1}{3^5}$, or $\dfrac{1}{243}$ **2.** a^{18} **3.** x^{12} **4.** 2^{21}

5. $\dfrac{5^3}{3^3}$, or $\dfrac{125}{27}$ **6.** 100 **7.** 1 **8.** $\dfrac{1}{x^2}$

9. y^{10} **10.** $7^6 x^{12}$ **11.** 4^{11} **12.** 3^4, or 81

13. $\dfrac{1}{t^{12}}$ **14.** $\dfrac{1}{z^2}$ **15.** $\dfrac{1}{2^5}$, or $\dfrac{1}{32}$ **16.** s^6

17. p **18.** 1 **19.** $\dfrac{y^{12}}{2^{20}}$ **20.** 10^{10}

21. w^9 **22.** c^7 **23.** $9x^4$ **24.** $\dfrac{1}{y^6}$

Integrated Review 23:   INTRODUCTION TO LOGARITHMS

Exercises

1. 2   **2.** 3   **3.** 1   **4.** 0   **5.** −1

6. 6   **7.** 1   **8.** 4   **9.** −2   **10.** −2

11. −5   **12.** 0

Answers: Preview Worksheets

Interactive Preview 1:   FUNCTION VALUES; DOMAIN AND RANGE

1.

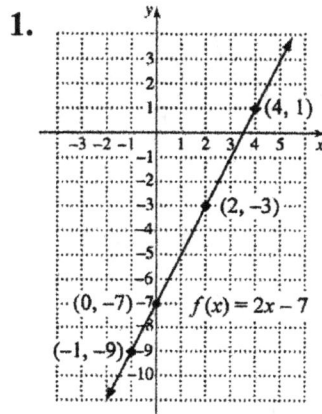

$f(2) = \underline{-3}$

$f(-1) = \underline{-9}$

$f(0) = \underline{-7}$

$f(4) = \underline{1}$

2.

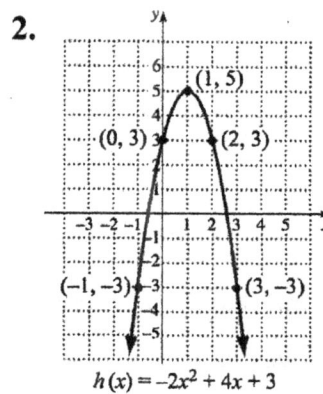

$h(0) = \underline{3}$

$h(1) = \underline{5}$

$h(-1) = \underline{-3}$

$h(3) = \underline{-3}$

$h(2) = \underline{3}$

3.

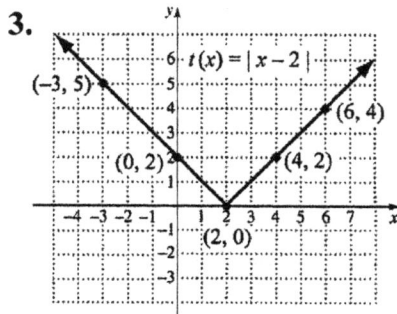

$t(-3) = \underline{5}$

$t(4) = \underline{2}$

$t(0) = \underline{2}$

$t(2) = \underline{0}$

$t(6) = \underline{4}$

4.

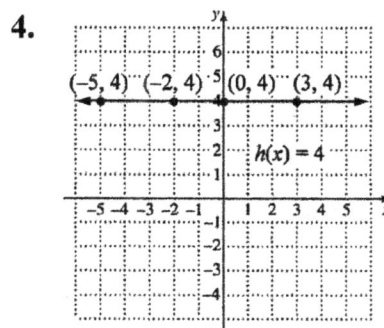

$h(3) = \underline{4}$

$h(-2) = \underline{4}$

$h(-5) = \underline{4}$

$h(0) = \underline{4}$

5.

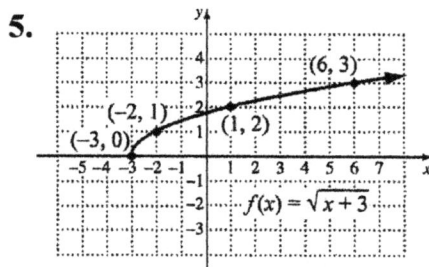

$f(-3) = \underline{0}$

$f(-2) = \underline{1}$

$f(1) = \underline{2}$

$f(6) = \underline{3}$

6.

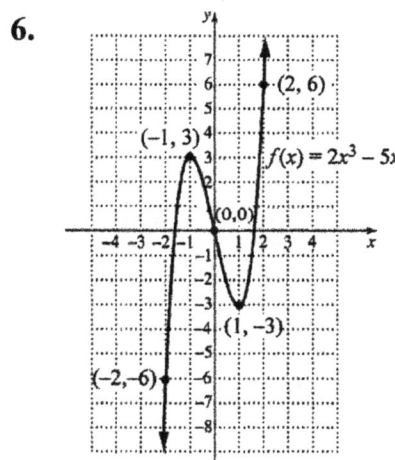

$f(-1) = \underline{3}$

$f(2) = \underline{6}$

$f(-2) = \underline{-6}$

$f(0) = \underline{0}$

$f(1) = \underline{-3}$

7. Domain: $(-\infty, \infty)$

  Range: $(-\infty, \infty)$

8. Domain: $(-\infty, \infty)$

  Range: $\{y \mid y \le 5\}$, or $(-\infty, 5]$

9. Domain: $(-\infty, \infty)$

  Range: $\{y \mid y \ge 0\}$, or $[0, \infty)$

10. Domain: $(-\infty, \infty)$

  Range: $\{4\}$

11. Domain: $\{x \mid x \ge -3\}$, or $[-3, \infty)$

  Range: $\{y \mid y \ge 0\}$, or $[0, \infty)$

12. Domain: $(-\infty, \infty)$

  Range: $(-\infty, \infty)$

Interactive Preview 2: GRAPHING PIECEWISE FUNCTIONS

1. Domain of Equation 1: $(-\infty, 2)$

   Domain of Equation 2: $[2, \infty)$

   Equation 1

$x < 2$	$f(x) = 3x$
1	3
0	0
−1	−3

   Equation 2

$x \geq 2$	$f(x) = x - 1$
2	1
4	3
5	4

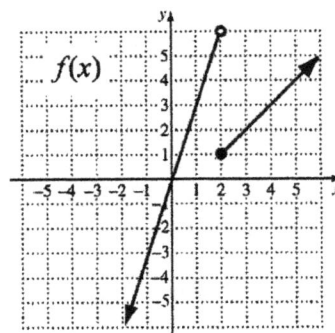

   - If $f(x) = 3x$ and $x = 2$, $f(2) = 3 \cdot 2 = 6$. Since 2 is not in the domain of $f(x) = 3x$, we use an open circle at $(2, 6)$.
   - If $f(x) = x - 1$ and $x = 2$, $f(2) = 2 - 1 = 1$. Since 2 is in the domain of $f(x) = x - 1$, we use a solid dot at $(2, 1)$.

2. Domain of Equation 1: $(-\infty, -1)$

   Domain of Equation 2: $[-1, 4)$

   Domain of Equation 3: $[4, \infty)$

   Equation 1

$x < -1$	$g(x) = -x + 1$
−2	3
−3	4
−4	5

   Equation 2

$-1 \leq x < 4$	$g(x) = -3$
−1	−3
1	−3
3	−3

   Equation 3

$x \geq 4$	$g(x) = 2x - 4$
4	4
5	6
6	8

   - If $g(x) = -x + 1$ and $x = -1$, $g(-1) = -(-1) + 1 = 2$. Since −1 is not in the domain of $g(x) = -x + 1$, we use an open circle at $(-1, 2)$.
   - If $g(x) = -3$ and $x = -1$, $g(-1) = -3$. Since −1 is in the domain of $g(x) = -3$, we use a solid dot at $(-1, -3)$.
   - If $g(x) = -3$ and $x = 4$, $g(4) = -3$. Since 4 is not in the domain of $g(x) = -3$, we use an open circle at $(4, -3)$.

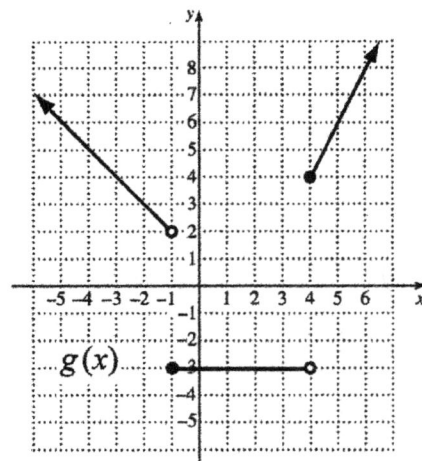

   - If $g(x) = 2x - 4$ and $x = 4$, $g(4) = 2 \cdot 4 - 4 = 4$. Since 4 is in the domain of $g(x) = 2x - 4$, we use a solid dot at $(4, 4)$.

3.

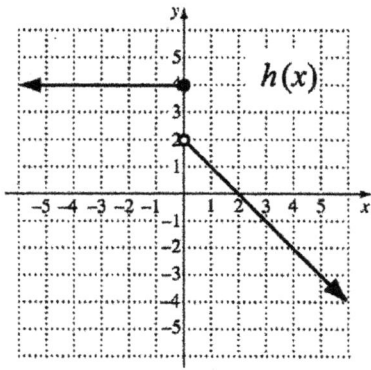

h(x)

4.

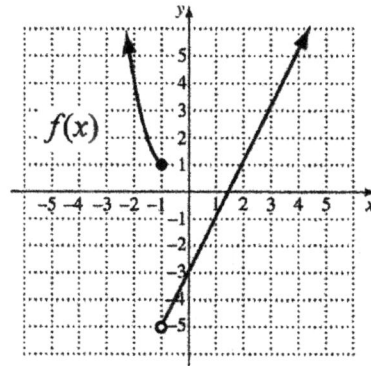

f(x)

5.

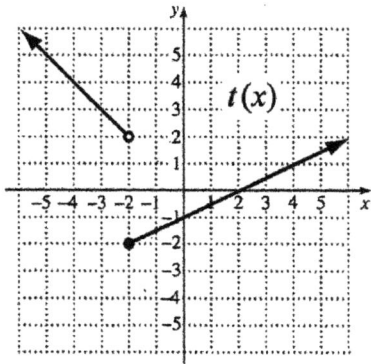

t(x)

6.

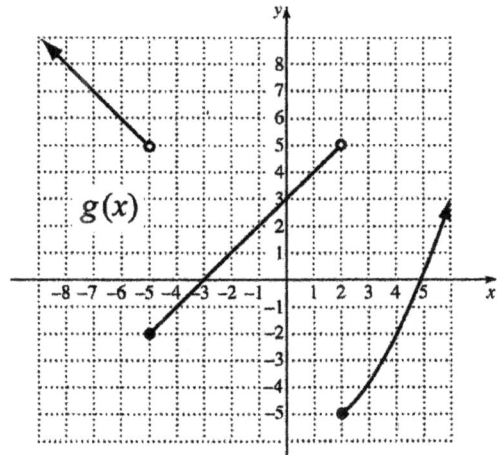

g(x)

Interactive Preview 3: TRANSFORMATIONS

1. (g)	2. (h)	3. (d)	4. (a)	5. (b)
6. (c)	7. (b)	8. (e)	9. (h)	10. (f)
11. (e)	12. (g)	13. (h)	14. (k)	15. (c)
16. (i)	17. (b)	18. (j)	19. (d)	20. (f)

1. $h(x) = x^2 - 8x + 15$

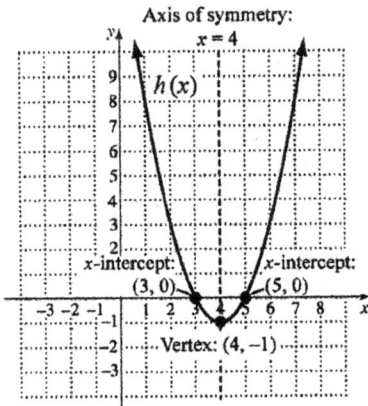

- $a = 1$; $a > 0$, thus the graph opens up.
- The zeros of $h(x)$ are 3 and 5.
- The minimum value of $h(x)$ is -1.

2. $f(x) = -x^2 - 4x - 3$

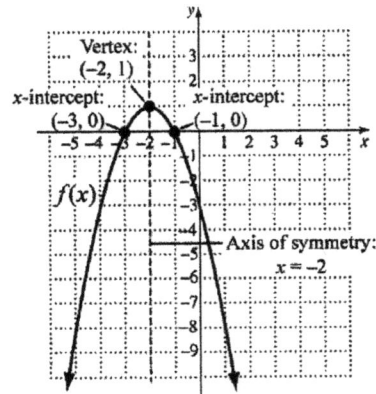

- $a = -1$; $a < 0$, thus the graph opens down.
- The zeros of $f(x)$ are -3 and -1.
- The maximum value of $f(x)$ is 1.

3. $g(x) = -x^2 - 2x + 3$

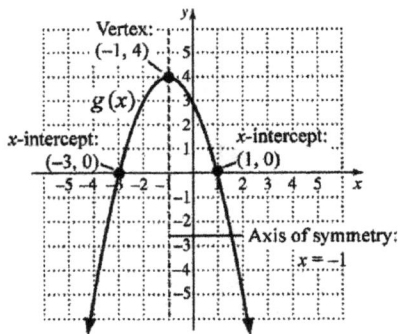

- $a = -1$; $a < 0$, thus the graph opens down.
- The zeros of $g(x)$ are -3 and 1.
- The maximum value of $g(x)$ is 4.

4. $f(x) = x^2 - 4x$

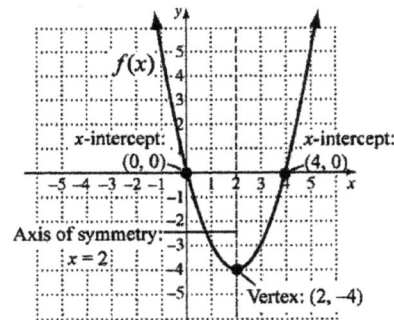

- $a = 1$; $a > 0$, thus the graph opens up.
- The zeros of $f(x)$ are 0 and 4.
- The minimum value of $f(x)$ is -4 .

Interactive Preview 5: SOLVING RATIONAL EQUATIONS

1. $\dfrac{40}{9}$ **2.** 13   **3.** -11 **4.** $-\dfrac{23}{7}$

5. $-\dfrac{2}{3}, 8$ **6.** $-2, -11$ **7.** -9 **8.** 5

Interactive Preview 6: CHECKING SOLUTIONS OF RADICAL EQUATIONS

Exercises

1. Yes   2. No   3. Yes   4. Yes

5. Only -1 is a solution.   6. Only 15 is a solution.

7. Only 1 is a solution.   8. Only 7 is a solution.

9. Both numbers are solutions.

Interactive Preview 7: SOLVING EQUATIONS AND INEQUALITIES WITH ABSOLUTE VALUE

1. 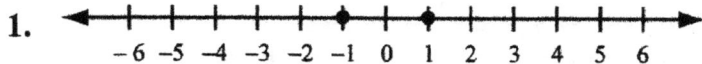 The solutions are -1 and 1.

2. 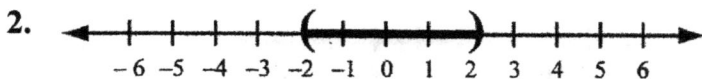 $(-2, 2)$

3. 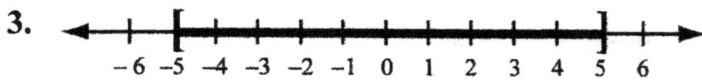 $[-5, 5]$

4. 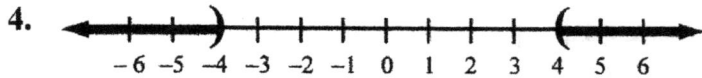 $(-\infty, -4) \cup (4, \infty)$

5. 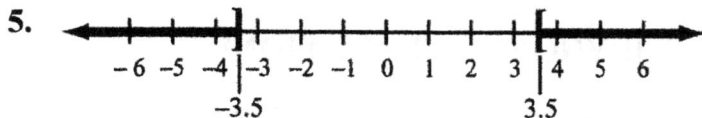 $(-\infty, -3.5] \cup [3.5, \infty)$

6. $q = -8 \ or \ q = 8$

7. $-2 \le p + 5 \le 2$

8. $0.5 - x < -6.5 \ or \ 0.5 - x > 6.5$

9. $7y - 1 = -12 \ or \ 7y - 1 = 12$

10. $-21 < \dfrac{3}{4}t + 6 < 21$

11. $18q \le -9 \ or \ 18q \ge 9$

12. $-30 \le w + 10 \le 30$
$-40 \le w \le 20$
Solution set: $[-40, 20]$

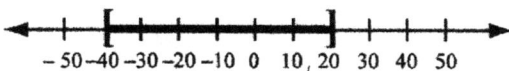

13. $y - 4 < -6 \ or \ y - 4 > 6$
$y < -2 \ or \ y > 10$
Solution set: $(-\infty, -2) \cup (10, \infty)$

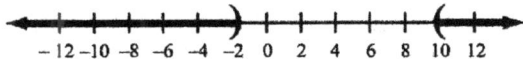

14. The solutions are -1 and 5.

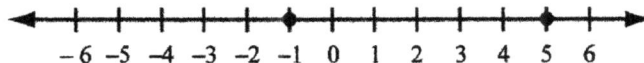

15. $(3, 5)$,

16. $(-\infty, -10] \cup [0, \infty)$,

17. The solutions are -1 and $-\dfrac{3}{5}$.

18. $(-\infty, -2) \cup (0, \infty)$,

19. $[-2, 3]$,

20. $[-1, 5]$,

21. $(-\infty, -2) \cup (6, \infty)$,

22. The solutions are -10 and 20.

23. $(-\infty, -2] \cup [1, \infty)$,

24. $(-8, 4)$,

25. $[-1, 7]$,

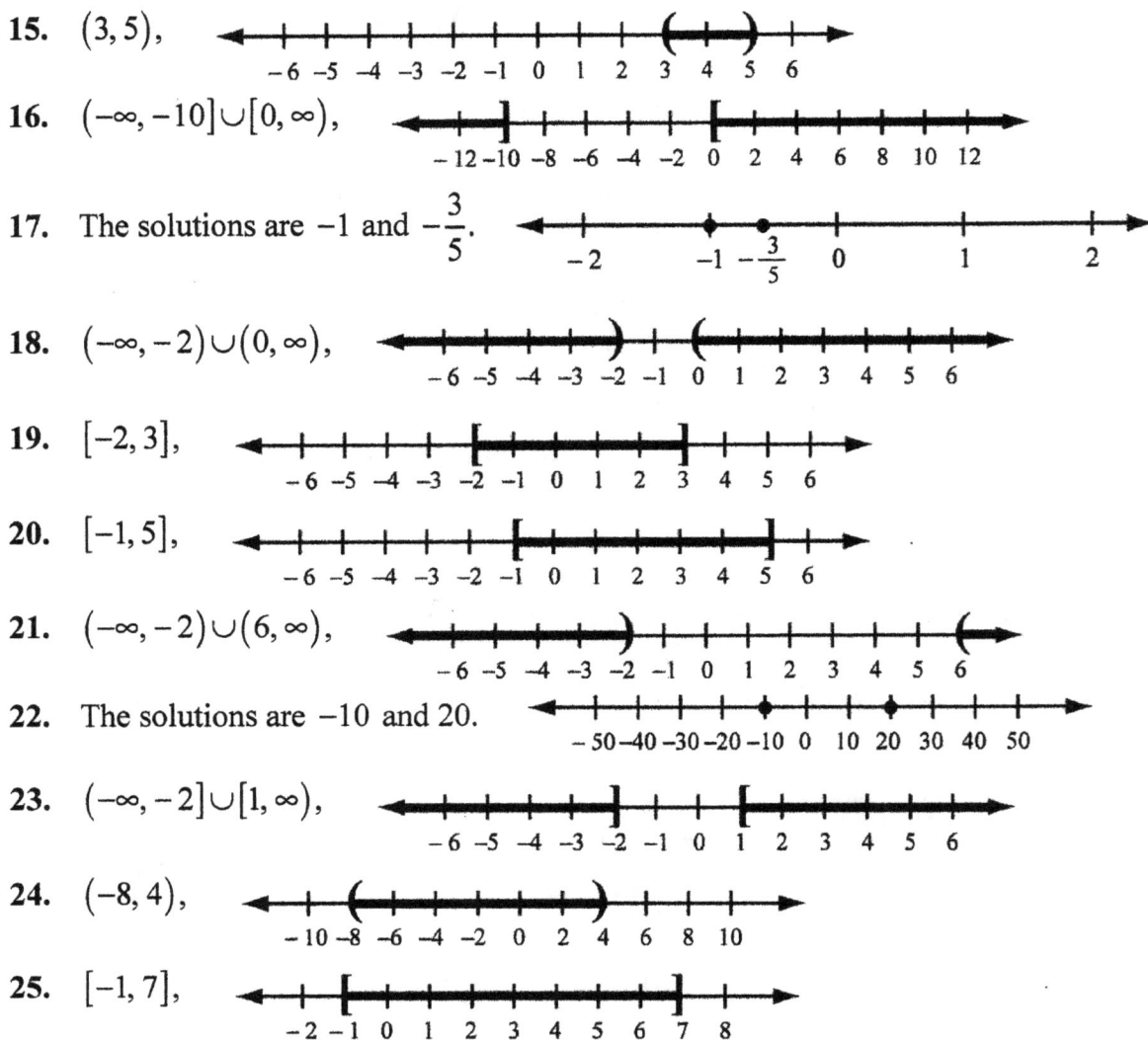

Interactive Preview 8:   ZEROS OF POLYNOMIAL FUNCTIONS

1. x-intercept of graph: $(1, 0)$

Solution of $3x - 3 = 0$: 1

Zero of $f(x) = 3x - 3$: 1

2. x-intercepts of graph: $(-1, 0)$ and $(4, 0)$

Solutions of $x^2 - 3x - 4 = 0$: -1 and 4

Zeros of $f(x) = x^2 - 3x - 4$: -1 and 4

3. x-intercepts of graph:
$(-2, 0)$, $(0, 0)$, and $(3, 0)$

Solutions of $x^3 - x^2 - 6x = 0$:
-2, 0, and 3

Zeros of $f(x) = x^3 - x^2 - 6x$:
-2, 0, and 3

4. $3x - 3 = 0$

$3x = 3$

$x = 1$

The zero of $f(x)$ is 1.

5. $(x+1)(x-4)=0$

$x+1=0$ or $x-4=0$

$x=-1$ or $x=4$

The zeros of $f(x)$ are -1 and 4.

6. $x(x+2)(x-3)=0$

$x=0$ or $x+2=0$ or $x-3=0$

$x=0$ or $x=-2$ or $x=3$

The zeros of $f(x)$ are 0, -2, and 3.

Interactive Preview 9: ASYMPTOTES OF RATIONAL FUNCTIONS

1. $f(x)=\dfrac{5x-1}{5x-10}$

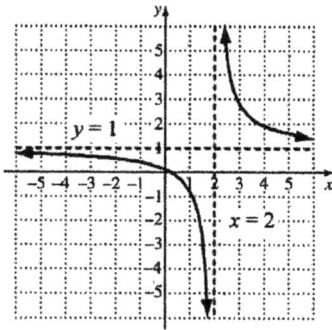

2. $f(x)=\dfrac{1}{x^2}$

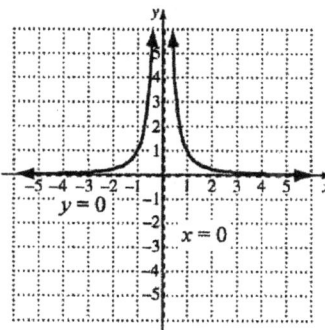

3. $f(x)=\dfrac{4}{x+2}$

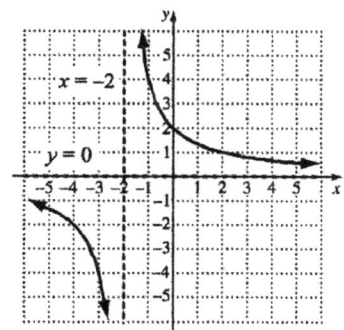

4. $f(x)=\dfrac{1+6x}{6+3x}$

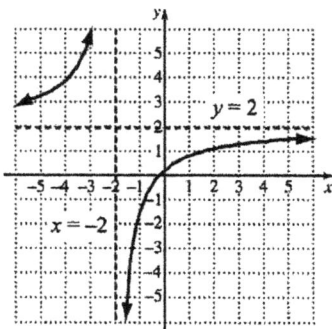

5. $f(x)=\dfrac{5}{x^2+3x}$

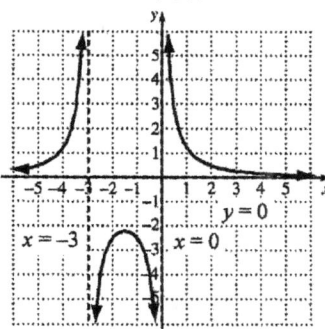

6. $f(x)=\dfrac{-2x^2+4x+3}{x^2-x-2}$

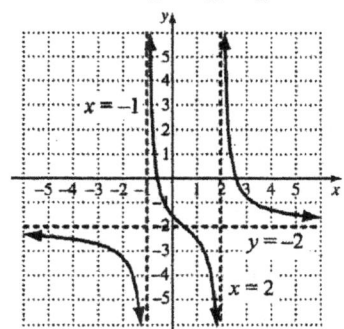

7. Zero of the denominator: 2; vertical asymptote: $x=2$

8. Zero of the denominator: 0; vertical asymptote: $x=0$

9. Zero of the denominator: -2; vertical asymptote: $x=-2$

10. Zero of the denominator: -2; vertical asymptote: $x=-2$

11. Zeros of the denominator: $-3, 0$; vertical asymptotes: $x=-3$, $x=0$

12. Zeros of the denominator: $-1, 2$; vertical asymptotes: $x=-1$, $x=2$

13. 1; 1; same as; 5; 5; $\dfrac{5}{5}$, or 1; $y=1$

14. 0; 2; less than; $y=0$

15. $y=0$ **16.** $y=2$ **17.** $y=0$ **18.** $y=-2$

Interactive Preview 10: GRAPHING INVERSE FUNCTIONS

1.

x	$f(x) = 2x + 4$
-4	-4
-3	-2
-1	2
0	4

x	$f^{-1}(x) = \dfrac{x-4}{2}$
-4	-4
-2	-3
2	-1
4	0

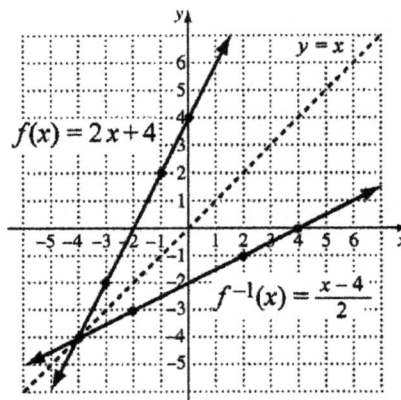

2.

x	$h(x) = 2 - 3x$
-1	5
0	2
1	-1
2	-4

x	$h^{-1}(x) = \dfrac{2-x}{3}$
5	-1
2	0
-1	1
-4	2

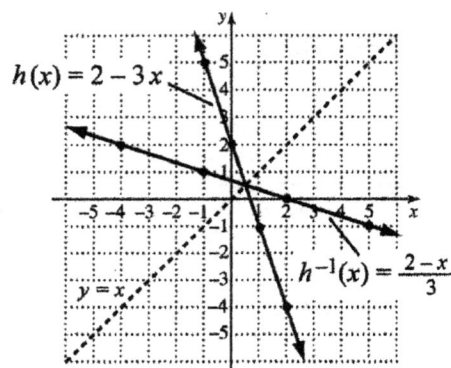

3.

x	$g(x) = x^3 - 3$
-2	-11
-1	-4
0	-3
1	-2
2	5

x	$g^{-1}(x) = \sqrt[3]{x+3}$
-11	-2
-4	-1
-3	0
-2	1
5	2

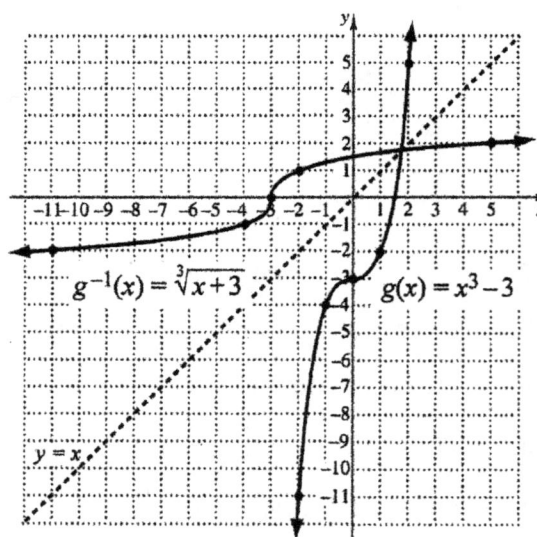

4.

x	$f(x)=x^2+4,\ x\geq 0$
0	4
1	5
2	8
3	13

x	$f^{-1}(x)=\sqrt{x-4}$
4	0
5	1
8	2
13	3

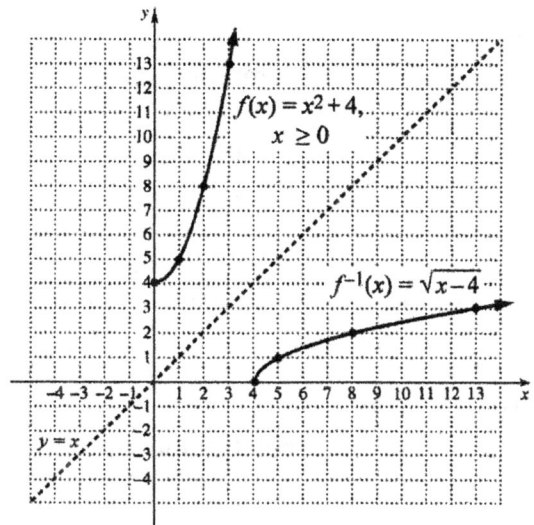

5.

x	$h(x)=x^2-3,\ x\geq 0$
0	−3
1	−2
2	1
3	6

x	$h^{-1}(x)=\sqrt{x+3}$
−3	0
−2	1
1	2
6	3

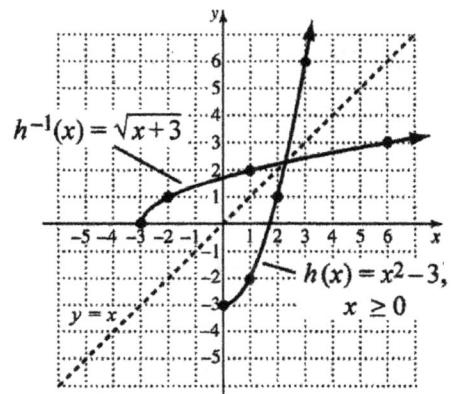

Interactive Preview 11: SOLVING EXPONENTIAL EQUATIONS

1.

$$5^x = 625$$

Write each side as a power of 5.	$5^x = 5^4$
Use the base-exponent property.	$x = 4$

2.

$$2^{3x+1} = 128$$

Write each side as a power of 2.	$2^{3x+1} = 2^7$
Use the base-exponent property.	$3x+1 = 7$
Subtract 1.	$3x = 6$
Divide by 3.	$x = 2$

3.
$$8^x = 25$$

Take the common logarithm on both sides.	$\log 8^x = \log 25$
Use the power rule.	$x \log 8 = \log 25$
Divide by $\log 8$.	$x = \dfrac{\log 25}{\log 8}$
Approximate using a calculator.	$x \approx 1.5480$

4.
$$e^x = 350$$

Take the natural logarithm on both sides.	$\ln e^x = \ln 350$
Use the power rule.	$x \ln e = \ln 350$
$\ln e = 1$	$x = \ln 350$
Approximate using a calculator.	$x \approx 5.8579$

5. 6   **6.** 1   **7.** 2.4849   **8.** $\dfrac{4}{3}$ **9.** 0.3155

10. 0.9575   **11.** 8   **12.** $\dfrac{1}{5}$ **13.** $\dfrac{3}{2}$ **14.** 1.6335

15. 0.8550   **16.** 2.0437   **17.** 4   **18.** 9.2103   **19.** 16

20. 69.3147   **21.** 0.2231   **22.** −3

Interactive Preview 12: SOLVING LOGARITHMIC EQUATIONS

1. 243   **2.** 1   **3.** −107   **4.** $\dfrac{23}{3}$

5. 100   **6.** 24   **7.** $\dfrac{1}{1000}$ **8.** 92

9. 2   **10.** 3   **11.** $\dfrac{13}{6}$ **12.** $\dfrac{5}{22}$

13. 13   **14.** 8   **15.** 25   **16.** $\sqrt{28}$, or $2\sqrt{7}$

Exercises

1. $\dfrac{4}{9}y = -16$

$\dfrac{9}{4} \cdot \dfrac{4}{9}y = \dfrac{9}{4}(-16)$

$1 \cdot y = -\dfrac{144}{4}$

$y = -36$

2. $-\dfrac{5}{2}t = -\dfrac{7}{10}$

$-\dfrac{2}{5}\left(-\dfrac{5}{2}t\right) = -\dfrac{2}{5}\left(-\dfrac{7}{10}\right)$

$1 \cdot t = \dfrac{14}{50}$

$t = \dfrac{7}{25}$

3.

$\begin{bmatrix} -3 & 4 \\ 5 & -7 \end{bmatrix} \cdot \begin{bmatrix} x \\ y \end{bmatrix} = \begin{bmatrix} -9 \\ 16 \end{bmatrix}$

$\begin{bmatrix} -7 & -4 \\ -5 & -3 \end{bmatrix} \cdot \begin{bmatrix} -3 & 4 \\ 5 & -7 \end{bmatrix} \begin{bmatrix} x \\ y \end{bmatrix} = \begin{bmatrix} -7 & -4 \\ -5 & -3 \end{bmatrix} \cdot \begin{bmatrix} -9 \\ 16 \end{bmatrix}$

$\begin{bmatrix} 1 & 0 \\ 0 & 1 \end{bmatrix} \cdot \begin{bmatrix} x \\ y \end{bmatrix} = \begin{bmatrix} -1 \\ -3 \end{bmatrix}$

$\begin{bmatrix} x \\ y \end{bmatrix} = \begin{bmatrix} -1 \\ -3 \end{bmatrix}$

The solution is $(-1, -3)$.

Check:

$\dfrac{-3x + 4y = -9}{-3(-1) + 4(-3) \mid -9}$

$3 - 12$

$\; -9 \mid -9 \quad \text{True}$

$\dfrac{5x - 7y = 16}{5(-1) - 7(-3) \mid 16}$

$-5 + 21$

$\; 16 \mid 16 \quad \text{True}$

4.

$\begin{bmatrix} 5 & -4 \\ 7 & -6 \end{bmatrix} \cdot \begin{bmatrix} a \\ b \end{bmatrix} = \begin{bmatrix} -6 \\ -10 \end{bmatrix}$

$\begin{bmatrix} 3 & -2 \\ \dfrac{7}{2} & -\dfrac{5}{2} \end{bmatrix} \cdot \begin{bmatrix} 5 & -4 \\ 7 & -6 \end{bmatrix} \begin{bmatrix} a \\ b \end{bmatrix} = \begin{bmatrix} 3 & -2 \\ \dfrac{7}{2} & -\dfrac{5}{2} \end{bmatrix} \cdot \begin{bmatrix} -6 \\ -10 \end{bmatrix}$

$\begin{bmatrix} 1 & 0 \\ 0 & 1 \end{bmatrix} \cdot \begin{bmatrix} a \\ b \end{bmatrix} = \begin{bmatrix} 2 \\ 4 \end{bmatrix}$

$\begin{bmatrix} a \\ b \end{bmatrix} = \begin{bmatrix} 2 \\ 4 \end{bmatrix}$

The solution is $(2, 4)$.

5. $-\dfrac{2}{11}$ **6.** -600 **7.** $(5, -2)$ **8.** $(5, -3)$

Interactive Preview 14: CLASSIFYING EQUATIONS OF CONIC SECTIONS

1. Parabola
2. Ellipse
3. Hyperbola
4. Hyperbola
5. Circle
6. Parabola
7. Parabola
8. Ellipse
9. Hyperbola
10. Circle
11. Parabola
12. Hyperbola
13. Ellipse
14. Circle
15. Ellipse
16. Parabola

Interactive Preview 15: THE ELLIPSE

1.

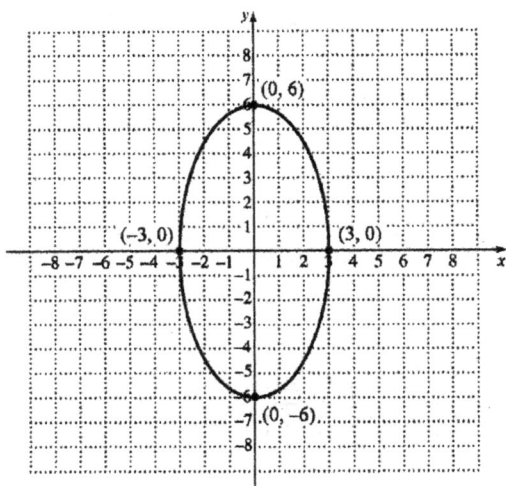

Center: $(0, 0)$

The major axis is vertical.
The minor axis is horizontal.

Vertices: $(0, -6)$ and $(0, 6)$

Endpoints of major axis: $(0, -6)$ and $(0, 6)$

Endpoints of minor axis: $(-3, 0)$ and $(3, 0)$

x-intercepts: $(-3, 0)$ and $(3, 0)$

y-intercepts: $(0, -6)$ and $(0, 6)$

Equation: $\dfrac{x^2}{3^2} + \dfrac{y^2}{6^2} = 1$, or $\dfrac{x^2}{9} + \dfrac{y^2}{36} = 1$

2.

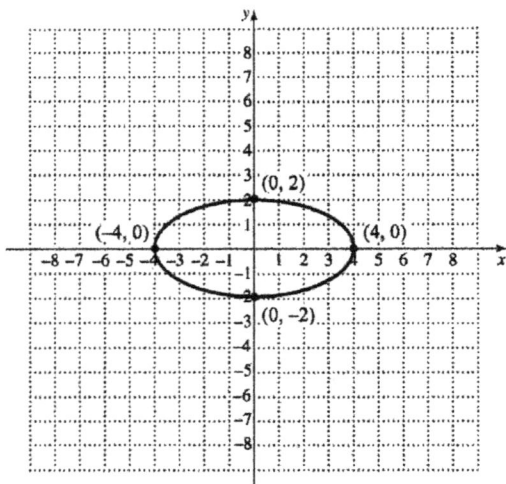

Center: $(0, 0)$

The major axis is horizontal.
The minor axis is vertical.

Vertices: $(-4, 0)$ and $(4, 0)$

Endpoints of major axis: $(-4, 0)$ and $(4, 0)$

Endpoints of minor axis: $(0, -2)$ and $(0, 2)$

x-intercepts: $(-4, 0)$ and $(4, 0)$

y-intercepts: $(0, -2)$ and $(0, 2)$

Equation: $\dfrac{x^2}{4^2} + \dfrac{y^2}{2^2} = 1$, or $\dfrac{x^2}{16} + \dfrac{y^2}{4} = 1$

Interactive Preview 16: THE HYPERBOLA

1.
   - The x^2-term is the first term, thus the transverse axis is on the x-axis.
   - The endpoints of the transverse axis are $(-6, 0)$ and $(6, 0)$.
   - The endpoints of the conjugate axis are $(0, -3)$ and $(0, 3)$.
   - The vertices are $(-6, 0)$ and $(6, 0)$.

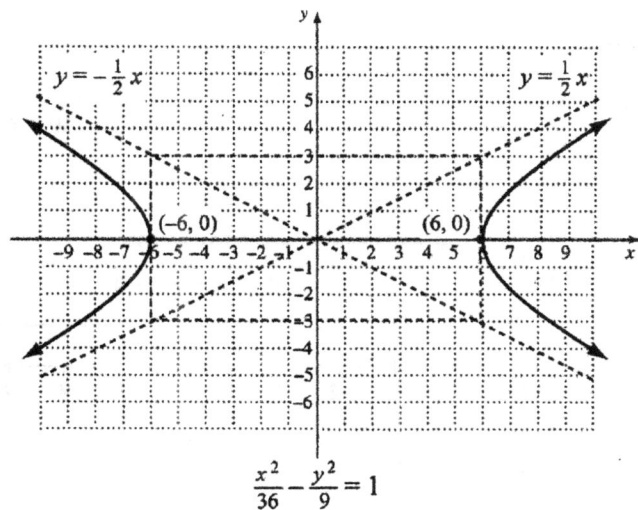

$$\frac{x^2}{36} - \frac{y^2}{9} = 1$$

2.
   - The y^2-term is the first term, thus the transverse axis is on the y-axis.
   - The endpoints of the transverse axis are $(0, -2)$ and $(0, 2)$.
   - The endpoints of the conjugate axis are $(-5, 0)$ and $(5, 0)$.
   - The vertices are $(0, -2)$ and $(0, 2)$.

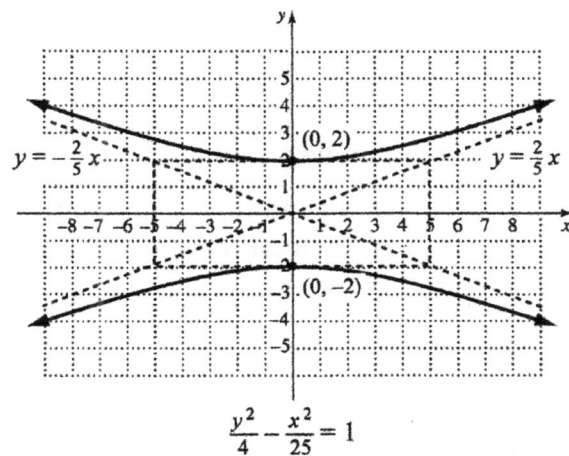

$$\frac{y^2}{4} - \frac{x^2}{25} = 1$$

3. E   4. D   5. A   6. B   7. F   8. C

Answers: Student Activities

Activity 1: American Football

1.

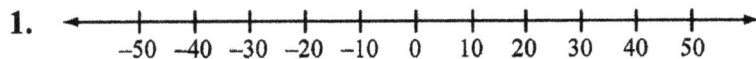

2.

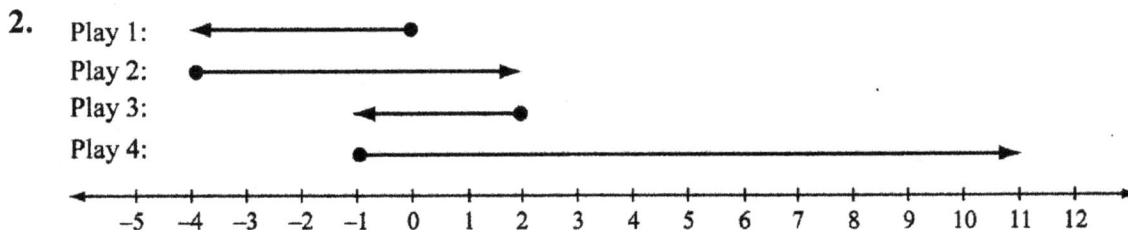

3. At the 11 yard mark on the number line

4. Yes

5. $-4 + 6 + (-3) + 12$

6. $-4 + 6 + 12 + (-3)$

7. The position was the same.

8. Yes; in the second set of plays, the ball would have moved more than 10 yards on the third down, so there would not have been a fourth down.

9. The direction is important, not just the distance.

10. $m = g - l$

11. Positive is good; 1 is a better turnover margin than -8

12.

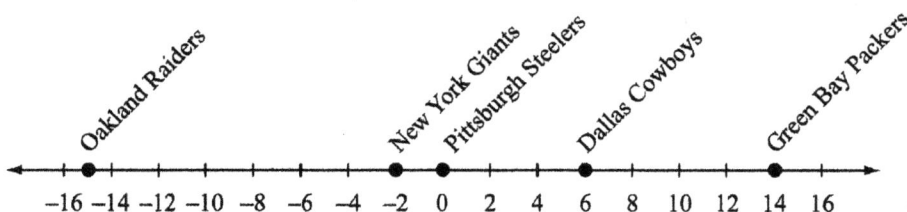

Activity 2: Finding the Magic Number

1. Jaguars' magic number with respect to the Catamounts: 5; Jaguars' magic number with respect to the Wildcats: 5

2. Standings will vary.

3. Magic numbers will vary.

4. Magic numbers will vary.

5. In the formula for the magic number, $G - P$ represents the number of games that the leading team will still play in the season and L represents how many more games the second-place team has lost than the leading team has lost. A magic number of 0 means that $G - P - L = -1$, so L is greater than $G - P$. Thus, there are not enough games remaining in the season for the leading team to lose as many games as the second-place team has lost. This indicates that the second-place team has been eliminated from contention.

Activity 3: Visualizing Factoring

1. (a) The student should have a total of one square and 13 rectangles.
   (b) Only the rectangle that is 1 unit by 10 units will work.
   (c) $x^2 + 11x + 10 = (x+1)(x+10)$; the dimensions of the rectangle are $x + 1$ and $x + 10$.

2. $x^2 + 7x + 10 = (x+2)(x+5)$.

3. 2 units by 6 units

Activity 4: Matching Factorizations

$x^2 - x - 6 = (x+2)(x-3)$ $x^2 + 5x + 6 = (x+2)(x+3)$

$2x^2 + 5x + 3 = (2x+3)(x+1)$ $x^2 - 5x + 6 = (x-2)(x-3)$

$6x^2 + x - 1 = (2x+1)(3x-1)$ $x^2 + x - 6 = (x-2)(x+3)$

$6x^2 + 5x - 6 = (2x+3)(3x-2)$ $6x^2 - 5x - 1 = (6x+1)(x-1)$

$6x^2 - 5x - 6 = (2x-3)(3x+2)$ $2x^2 - 5x + 3 = (2x-3)(x-1)$

$6x^2 - 5x + 1 = (2x-1)(3x-1)$ $6x^2 + 13x + 6 = (2x+3)(3x+2)$

$6x^2 + 5x - 1 = (6x-1)(x+1)$ $2x^2 + x - 3 = (2x+3)(x-1)$

$6x^2 - x - 1 = (2x-1)(3x+1)$ $6x^2 - 7x + 1 = (6x-1)(x-1)$

Activity 5: Pets in the United States

1. The number of households with one or more of the pets in 2007 and in 2012 and total pet population in 2007 and in 2012 are given; each number should be multiplied by 1000; 43,021,000 households had dogs in 2007.

2. Number of households increased for dogs, turtles, snakes, and lizards; total pet population increased for turtles, snakes, and lizards; the number of pets per household changed.

3. $a = \dfrac{p}{h}$; cats, birds, horses, snakes

4. Fill-in answers are shown in **bold**.

	Households (in thousands) in 2007	Households (in thousands) in 2012	Change in the Number of Households (in thousands)	Percent Change in the Number of Households
Dogs	43,021	43,346	+325	+ 0.8%
Cats	37,460	36,117	−1,343	−3.6%
Birds	4,453	3,671	**−782**	**−17.6%**
Horses	2,087	1,780	**−307**	**−14.7%**
Fish	9,036	**7,738**	−1,298	**−14.4%**
Ferrets	**505**	334	−171	**−33.9%**
Rabbits	1,870	**1,408**	**−462**	−24.7%
Hamsters	**826**	877	**+51**	+6.2%
Guinea Pigs	**628**	847	+219	**+34.9%**
Gerbils	187	**234**	**+47**	+25.1%
Turtles	1,106	1,320	**+214**	**+19.3%**
Snakes	390	555	**+165**	**+42.3%**
Lizards	719	726	**+7**	**+1.0%**

5. The greatest increase was in the number of households containing dogs; the greatest percent increase was in the number of households containing snakes; there were many more households containing dogs than snakes in 2007, so a large increase in the number of households containing dogs did not result in a large percent increase.

6. In the 5 years from 2007 to 2012, the number of households containing gerbils rose by 47,000. In another 5 years (2017), if the trend remains the same, there will be 234,000 + 47,000, or 281,000 households containing gerbils. In the 5 years from 2007 to 2012, the number of households containing gerbils rose by 25.1%. In another 5 years (2017), if the trend remains the same, there will be $234,000(1.251) = 292,734$ households containing gerbils. Assuming that the percent change remained the same resulted in the larger estimate, because the percent change was calculated on a larger amount (234,000 rather than 187,000).

7. 11,500,000 fish

8. 1.3 hamsters per household

9. Answers will vary.

Activity 6: Linear Regression on a Graphing Calculator

1. The coefficient a represents the slope of the regression line. This is the rate of change in TV viewing in the given age group. The constant b represents the y-intercept of the regression line.

2. The line will slant down from left to right because the slope is negative. The graph confirms this.

3. 2019: 14.82%; 2020: 12.46%

4. About 11 years after 2014, or in 2025; answers regarding whether this seems reasonable will vary.

5. $y = 0.44x + 17.44$

6. 2020: \$19.64 billion; 2021: \$20.08 billion

7. Answers will vary.

8. Answers will vary.

Activity 7: Going Beyond High School

1. Each number represents a percent of high school completers who enrolled in a college; the column heads reference "high school completers" because some students have a High School Equivalency instead of having graduated from high school with a diploma.

2. 1,961,843; 708,526; 1,253,317; 1,015,157

3. Graphs may vary; a circle graph may be the best choice.

4. Graphs may vary; a bar graph may be the best choice.

5. Graphs may vary; a line graph may be the best choice. The graphs will show that, overall, enrollment in two-year colleges and in four-year colleges is increasing.

6. 0.2868% per year; 0.5105% per year; on average, the percent of male (or female) high school completers increased by the given amount each year; answers may vary.

7. Males: $y = 0.2868x + 52.6$, Females: $y = 0.5105x + 49.0$, where x = the number of years after 1975 and y = the percent enrolling in college; answers may vary.

8. Males: $y = 0.3787x + 51.5331$, Females: $y = 0.6074x + 50.2203$, where x = the number of years after 1975, y = the percent enrolling in college, and coefficients are rounded to four decimal places; answers may vary.

9. The third and fourth columns of the first table are further broken down into those attending two- and four-year colleges.

10. Answers may vary.

11. Answers may vary.

Activity 8: Data and Downloading

1. As the file size increases, the download time increases and the number of downloads decreases.

2. The common ratio is 0.1; this represents 0.1 sec/MB, the time it takes to download one MB of data.

3. The common product is 5000; this represents 5000 MB, or 5 GB, the monthly plan limit.

4.

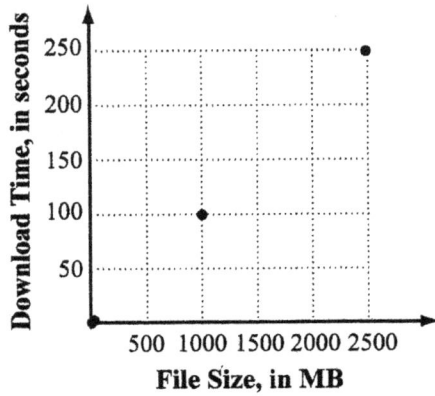

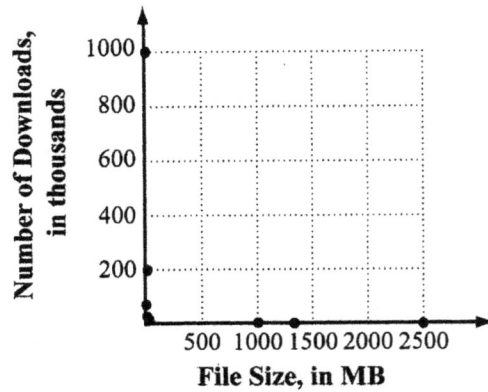

Note: In this graph, there are four points that lie very close to $(0, 0)$.

5. File size and download time vary directly; file size and number of downloads vary inversely.

6. $t = 0.1x$, where t is in seconds and x is in MB; the variation constant is the common ratio.

7. $n = \dfrac{5000}{x}$, where n is the number of downloads and x is in MB; the variation constant is the common product.

8. $t = 0.1(2) = 0.2$ seconds; $n = 5000 / 2 = 2500$ songs

9. Answers may vary.

Activity 9: Mobile Data

1. No; the relative positions of Central and Eastern Europe, Middle East and Africa, Western Europe, and Latin America have changed.

2. Total row for 2016 to 2019: 6,766; 10,666; 16,140; 24,220
Amount of Increase column missing data: 2,852; 2,051; 1,831; 21,697
Percent Increase column missing data: 1341%; 1433%; 601%; 911%; 860%.
Asia Pacific saw the greatest amount of increase, and Middle East and Africa saw the greatest percent increase.

3. **(a)** $W(2019) = 2,392$; this represents the average monthly mobile data usage in Western Europe in 2019

(b) $(C + W)(2014) = 583$; this represents the average monthly mobile data usage in Central, Eastern, and Western Europe in 2014

(c) $(N - L)(2019) = 1,766$; this represents how much greater the average monthly mobile data usage was in North American than in Latin America in 2019

(d) $(A + N + C + M + W + L)(2019) = 24,220$; this represents the total average monthly mobile data usage in 2019 and is the same as $T(2019)$.

4. The heights of the regions represent the regional monthly mobile data functions.

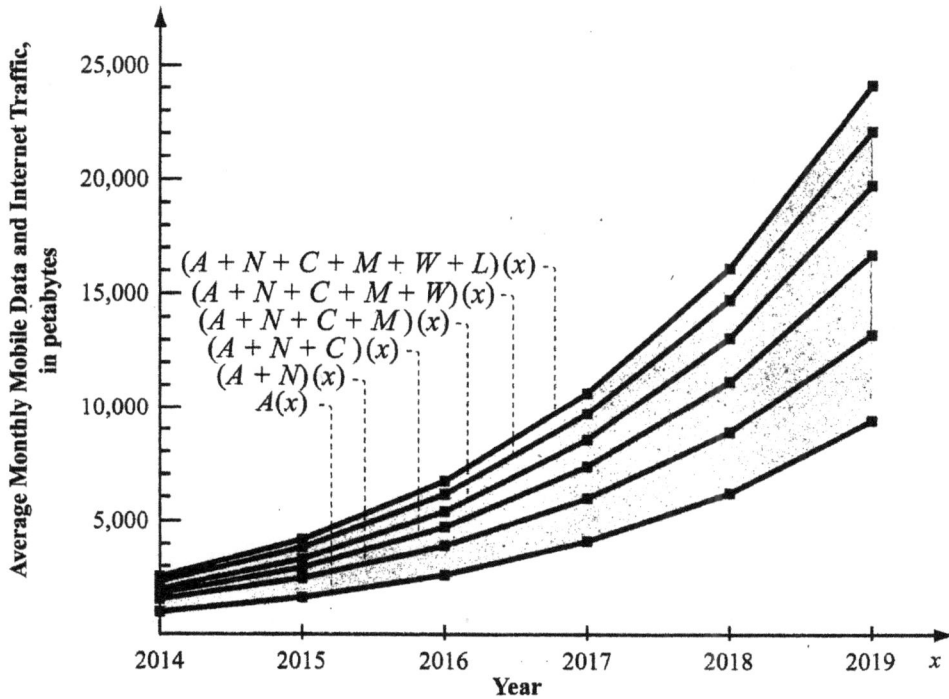

5. Answers may vary.

Activity 10: Firefighting Formulas

1. Increase; answers may vary.

2., 3.

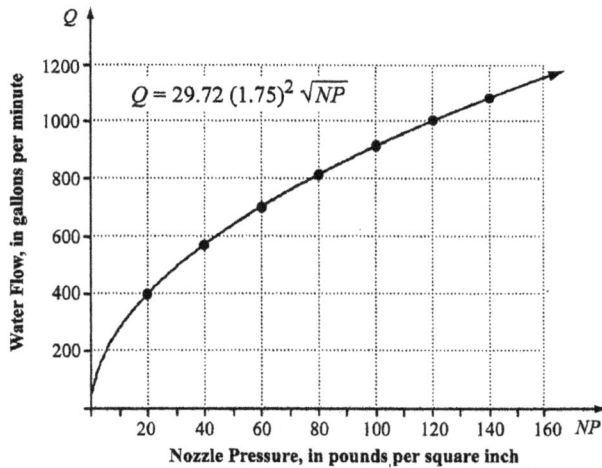

4. $NR = 0.05Q \times \sqrt{NP}$

5. $Q = 518$ gallons per minute; $NR = 212$ pounds; $V = 94$ feet per second

6. 96, 192, 288, 385, 481, 577, 673

7. Answers may vary.

Activity 11: Music Downloads

1. 428,000,000 albums; 257,000,000 albums; 1,070,000,000 single tracks; 1,108,000,000 single tracks

2. Decrease; increase

3. Overall, there was a decreasing trend in the sales of digital albums; and a linear function appears to be a good choice to model the data. Answers may vary.

4. The sales of digital single tracks increased from 2008 to 2012 and then began to decrease. A quadratic function might be a good choice to model the data. Answers may vary.

5. Answers may vary.

6. Answers may vary.

7. $A(x) = -25.1x + 407.6$, where $A(x)$ is the number of digital albums sold, in millions, x years after 2008 and coefficients are rounded to the nearest tenth; answers may vary.

8. $T(x) = -20.5x^2 + 139.8x + 1043.0$, where $T(x)$ is the number of digital single tracks sold, in millions, x years after 2008 and coefficients are rounded to the nearest tenth; answers may vary.

9. Yes; using the functions from Questions 7 and 8, sales will be the same 10.9 years after 2008, or in about 2019; answers may vary.

10. Sales of single tracks were a maximum in 2012. Answers may vary.

Activity 12: Let's Go to the Movies

1. Answers may vary.

2. Answers may vary.

3. Answers may vary; function found by regression is $E(x) = -0.0026x^2 + 0.0597x + 1.1252$, where x is the number of years after 1992 and coefficients are rounded to 4 decimal places.

4. Answers may vary; linear function found by regression is $F(x) = 0.0415x + 1.1341$, where x is the number of years after 1992 and coefficients are rounded to 4 decimal places.

5. Answers may vary; quadratic function found by regression is $S(x) = 0.0038x^2 - 0.1371x + 2.565$, where x is the number of years after 1992 and coefficients are rounded to 4 decimal places.

6. No; using the functions above, E was the closest; answers may vary.

7. Answers may vary.

8. Answers may vary.

Activity 13: Teaching About Zeros: Creating a Video

1-4. Answers will vary.

Activity 14: Earthquake Magnitude

1. Strongest: September 16, 2015; weakest: March 18, 2015. Answers may vary.

2. $M_0 = 10^{\frac{3}{2}(M_W + 10.7)}$; $3.5 \times 10^{26}, 2.2 \times 10^{28}, 2.0 \times 10^{27}, 3.5 \times 10^{26}, 4.0 \times 10^{27}, 4.5 \times 10^{25}$, $3.5 \times 10^{26}, 3.2 \times 10^{25}, 4.5 \times 10^{25}, 3.2 \times 10^{28}$

3. 1000 times stronger

4. Yes

5. Let m = the moment of magnitude of the weaker earthquake. Then $m + 1$ = the moment of magnitude of the stronger earthquake. Divide the seismic moments to determine how many times stronger the stronger earthquake is than the weaker earthquake:

$$\frac{10^{\frac{3}{2}(m+1+10.7)}}{10^{\frac{3}{2}(m+10.7)}} = 10^{\frac{3}{2}(m+1+10.7) - \frac{3}{2}(m+10.7)} = 10^{\frac{3}{2}m + \frac{3}{2} + \frac{3}{2}(10.7) - \frac{3}{2}m - \frac{3}{2}(10.7)} = 10^{\frac{3}{2}}.$$

So the stronger earthquake is $10^{\frac{3}{2}}$ or about 31.6 times as strong as the weaker earthquake.

6. $3.5 \times 10^{17}, 1.1 \times 10^{19}, 3.5 \times 10^{20}, 1.1 \times 10^{22}, 3.5 \times 10^{23}, 1.1 \times 10^{25}, 3.5 \times 10^{26}, 1.1 \times 10^{28}$, 3.5×10^{29}. Answers may vary.

7. The relative magnitudes are easier to represent and discuss. Answers may vary.

Activity 15: Fruit Juice Consumption

1. Answers may vary.

2. Answers may vary.

3. Answers may vary; using linear regression and all the data points, letting x represent the number of years after 1985, g represent grape juice consumption, f represent grapefruit juice consumption, and p represent pineapple juice consumption, and rounding coefficients to 3 decimal places:
   $g = 0.006x + 0.279; \quad f = -0.022x + 0.797; \quad p = -0.008x + 0.432$

4. Answers may vary; g is increasing, and f and p are decreasing.

5. g and f intersect at about (18, 0.4); this means that grape juice consumption and grapefruit juice consumption were the same (0.4 gallons per person) 18 years after 1985, or in 2003. g and p intersect at about (11, 0.3); this means that grape juice consumption and pineapple juice consumption were the same (0.3 gallons per person) 11 years after 1985, or in 1996. f and p intersect at about (25, 0.2); this means that grapefruit juice consumption and pineapple juice consumption were the same (0.2 gallons per person) 25 years after 1985, or in 2010.

6. Answers may vary; using linear regression and all the data points, letting x represent the number of years after 1985, r represent orange juice consumption, and a represent apple juice and cider consumption, and rounding coefficients to 3 decimal places: $r = -0.014x + 4.573$; $a = 0.015x + 1.573$. Predicted orange juice consumption in 2015 is 4.15 gallons per person; predicted apple juice and cider consumption in 2015 is 2.02 gallons per person. Orange juice consumption will equal apple juice and cider consumption in 2088.

7. Answers may vary.

8. Answers may vary.

9. Answers may vary.

10. Using linear regression and all the data points, letting x represent the number of years after 2000, r represent orange juice consumption, and a represent apple juice and cider consumption, and rounding coefficients to 3 decimal places: $r = -0.178x + 5.359$; $a = 0.045x + 1.799$. Predicted orange juice consumption in 2015 is 2.70 gallons per person; predicted apple juice and cider consumption in 2015 is 2.47 gallons per person. Orange juice consumption will equal apple juice and cider consumption in 2016.

11. Answers may vary.

Activity 16: Waterfalls

1. Fan: $c \le \frac{1}{3}b$;   Funnel: $c > 3b$;

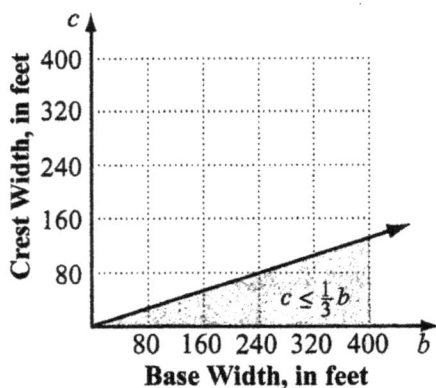
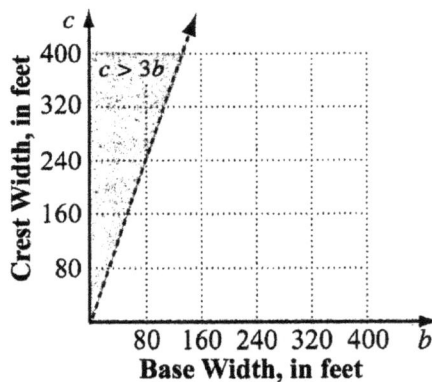

Curtain: $c > \dfrac{3}{2}h$;

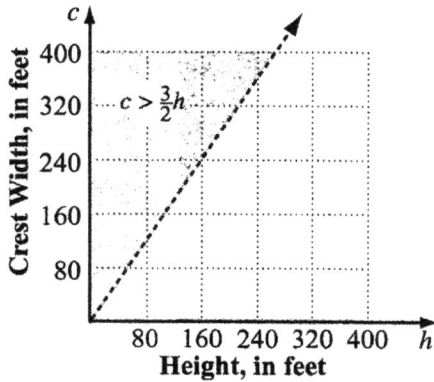

Ribbon: $c \le \dfrac{1}{2}h$;

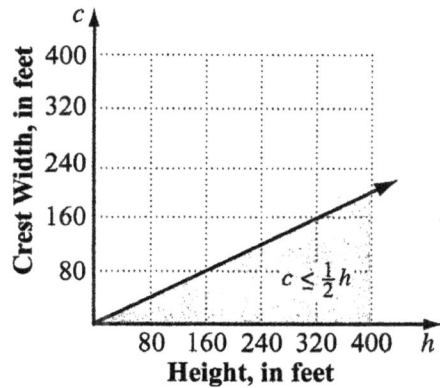

2. (a) Ribbon; **(b)** fan and ribbon; **(c)** curtain

Activity 17: Construction

1. **(a)** and **(d)**

2. $h \ge 30$

3. $d + h \ge 48$

4. Answers may vary.

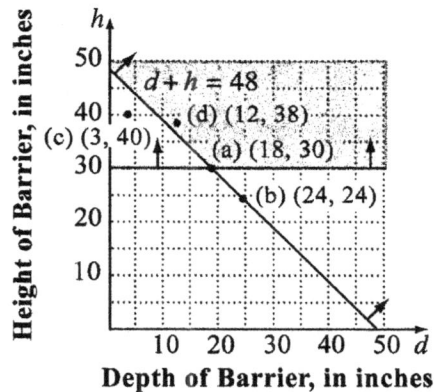

Activity 18: Cosmic Path

1. 3.618414 AU; 337 million mi

2. $b^2 = a^2 - (a - 2.053218)^2$; $b \approx 3.262374$ AU; $b^2 = a^2 - (a - 191)^2$; $b \approx 304$ million mi

3. 21.631105 AU; 2012 million mi

4. 3.144056 AU/yr; 33,400 mph

5. Answers may vary.

Activity 19: Bargaining for a Used Car

1. $2833.33

2. $1000 or less

3. $2833.33; $1000 or less

4. $P = 5500 - \dfrac{2}{3}d$

www.ingramcontent.com/pod-product-compliance
Lightning Source LLC
Chambersburg PA
CBHW061408210326
41598CB00035B/6142